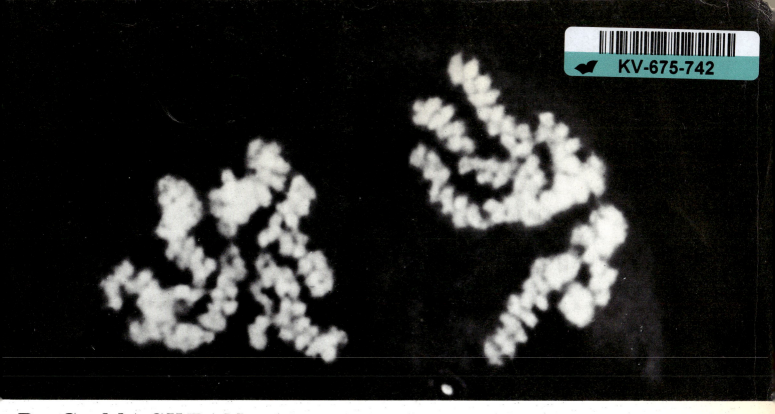

D. G. MACKEAN *BA FI Biol. Sir Frederic Osborn School, Welwyn Garden City*
Introduction to Genetics

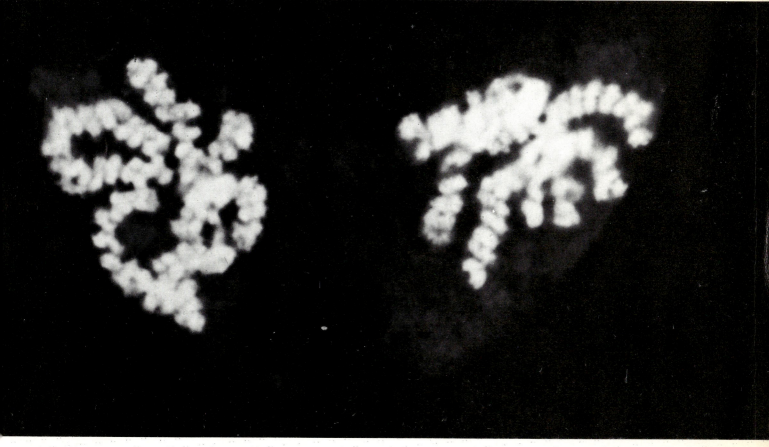

JOHN MURRAY   ALBEMARLE STREET   LONDON

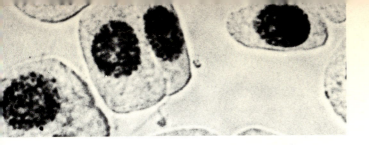

Preface

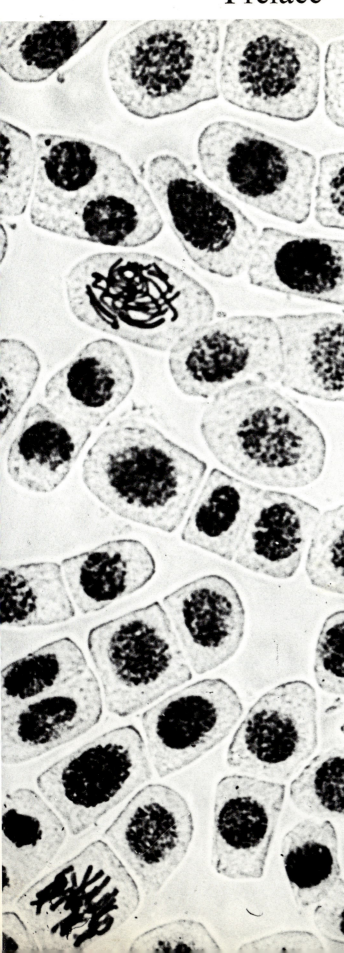

THIS book is intended for students in the fourth, fifth or lower sixth forms of secondary schools who are following or have completed an O-level biology course. The material included is suitable for Human Biology or Anatomy, Physiology and Hygiene syllabuses.

A heuristic approach has not been attempted, the classical Mendelian experiments have not been described and, in fact, Mendel himself has been relegated to a footnote. In explaining the principles of Mendelian inheritance, full use has been made of modern knowledge of cytology, to which the first chapter is largely devoted. This has led to the introduction of Mendelian terminology before the terms have been defined, but it is hoped that the cross references will enable the reader to acquire an understanding of these terms sufficient to follow the subsequent explanations.

The chapter on Evolution and Natural Selection has been included to relate the understanding of heritable variation to its long-term importance in producing change. As is inevitable in an introductory work, however, most of the space is devoted to explaining evolution and natural selection as concepts.

September 1967   D.G.M.

In the second edition, Chapter 3 has been extended so as to show the interrelationship between genetics and natural selection.

March 1971   D.G.M.

The third edition includes a more detailed account of DNA replication and transcription and other topics such as "back-crossing" and "continuous and discontinuous variation" have been introduced. The opportunity has also been taken to improve the accuracy in the use of some of the terminology.

January 1977   D.G.M.

Acknowledgements

The author is indebted to Dr C. O. Carter, Dr R. F. Kemp and Mr S. W. Hurry for reading the manuscript and for their many helpful suggestions and corrections in both first and second editions. Grateful thanks are also due to J. McLeish and B. Snoad, Dr A. H. Sparrow, Professor Wolfgang Beermann and the World Health Organization for the photographic plates.

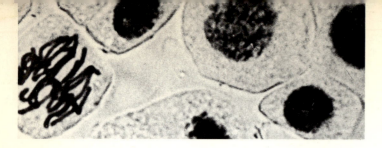

Contents

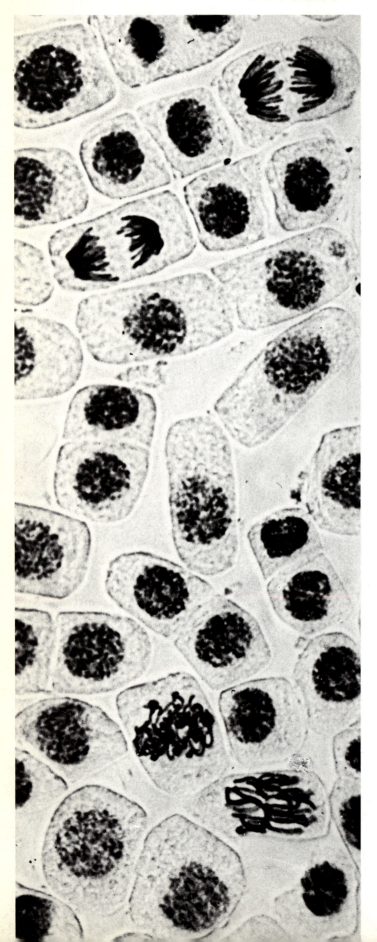

Glossary inside back flap
The drawings are by the author

1 | Chromosomes and Heredity

GENETICS is the study of the mechanisms by which character-istics of parent organisms are passed on to their offspring. These offspring are usually the products of single cells resulting from the fusion of the parental reproductive cells, called gametes. It follows, therefore, that whatever the mechanism of inheritance may be, it must reside in the cell, and therefore an examination of the cell and its method of division may provide clues about the way in which hereditary 'instructions' are passed on.

Cell division

In the early stages of growth and development of an organ-ism, all the cells are actively dividing to produce new tissues and organs. Later, particularly when the cells become specialized, this power of division is lost and only a limited number of unspecialized cells retain the power of division, e.g. cambium cells in plants and cells of the Malpighian layer in the skin which produce new epidermis. In those cells that continue to divide, the sequence of events leading to cell division is basically the same. Firstly, the nucleus divides into two and then the whole cell divides, separating each nucleus in a unit of cytoplasm, so that two cells now exist where previously there was only one. Both cells may then enlarge to the size of the parent cell. Such cell division and enlargement gives rise to growth.

The detailed sequence of events which takes place when the nucleus of a cell divides has been worked out over the last eighty years. A fairly close study of these events reveals the way in which inherited characters can, in the course of cell division, be passed on to the new cells produced and thus, eventually, to the new organism formed from these cells.

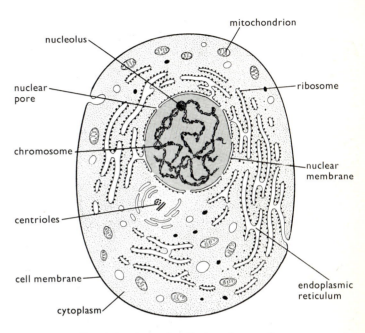

Fig. 1.1   Animal cell at early prophase

Mitosis (Fig. 1.2 and Plate 1)

Prior to division, the nucleus of the cell enlarges and in the nucleus there appear a definite number of fine, coiled, thread-like structures called *chromosomes* (Fig. 1.1). The behaviour of these chromosomes during cell division is called mitosis, which

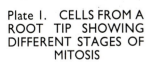

Plate I. CELLS FROM A ROOT TIP SHOWING DIFFERENT STAGES OF MITOSIS

Magnification approx. ×500

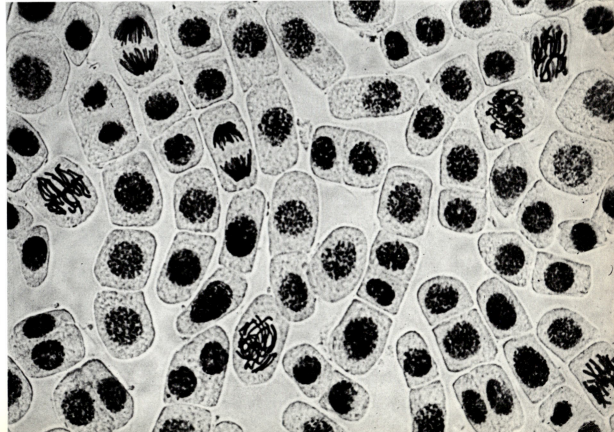

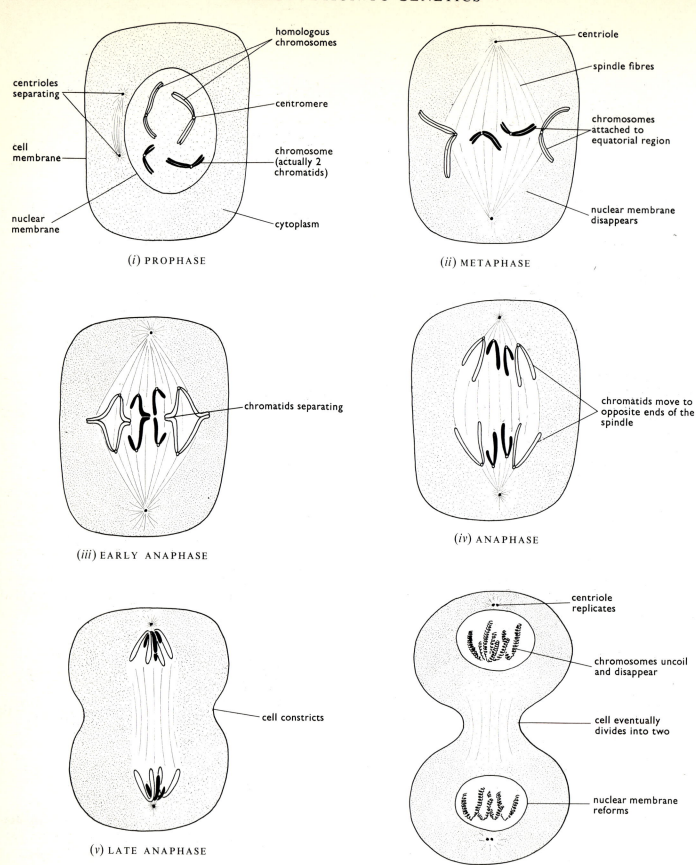

Fig. 1.2(a)   Mitosis in an animal cell

Although the stages of mitosis are necessarily shown as static events, it must be emphasized that the process is a continuous one and the names 'Anaphase', 'Metaphase', etc., do not imply that the process of mitosis comes to a halt at this juncture. Moreover, the stages shown are not selected at regular intervals of time, e.g. in the embryonic cells of a particular grasshopper the timing at 38°C is as follows:

Prophase 100 mins   **Metaphase** 15 mins   **Anaphase** 10 mins   **Telophase** 60 mins

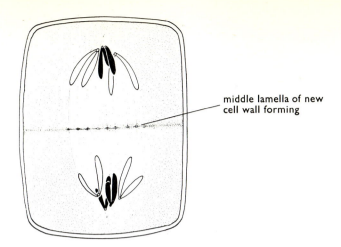

Fig. 1.2(*b*) Late anaphase in a plant cell showing how separation of daughter cells differs from animal cells

is usually described as a series of stages, prophase, metaphase, etc., though, in fact, the events occur in a smoothly continuous pattern and do not occupy equal periods of time (*see* p. 2).

1. **Prophase.** The chromosomes become more pronounced (that is, they react more readily to stains and chemical fixatives). They shorten and thicken (Plate 3*a*), probably by coiling like a helical spring, but with the coils so close to each other that they are not visible at low magnifications (Fig. 1.3 and Plate 2). The nuclear membrane dissolves, leaving the chromosomes suspended in the cytoplasm, and at the same time the one or more *nucleoli* disappear.

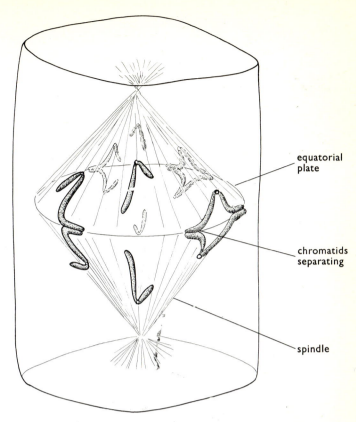

Fig. 1.4 Stereogram of cell at anaphase

2. **Metaphase.** In the cells of animals and some of the simpler plants there is a pair of minute bodies called the *centrioles* which lie just outside the nucleus. At this stage they move away from each other and migrate to opposite ends of the cell. From each centriole there radiate what appear to be cytoplasmic fibres which meet and join near the centre of the cell. This system of 'fibres' makes a web-like structure called the *spindle* (Fig. 1.4) and the chromosomes become attached by their centromeres (Fig. 1.3) to the equatorial region of the spindle. (Most plant cells do not have centrioles but a spindle is

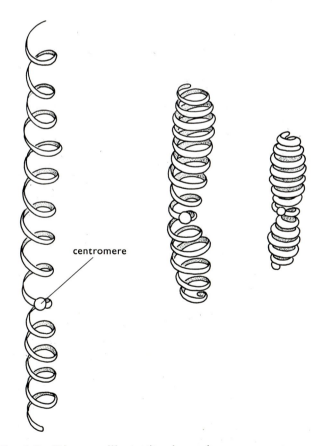

Fig. 1.3 Diagram illustrating how chromosomes appear to become thicker and shorter during prophase

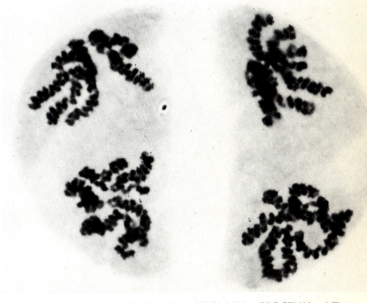

Plate 2. CHROMOSOMES OF *TRILLIUM ERECTUM* AT TELOPHASE OF MEIOSIS (p. 11), SHOWING COILING
(A. H. Sparrow, Ph.D., Brookhaven National Laboratory, U.S.A.)

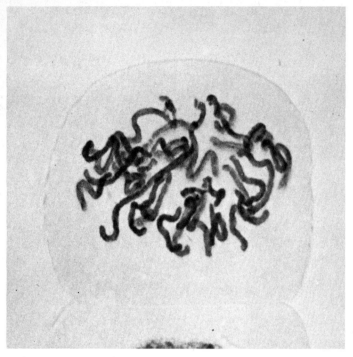

Plate 3*a*.   PROPHASE
The chromosomes have become short and thick

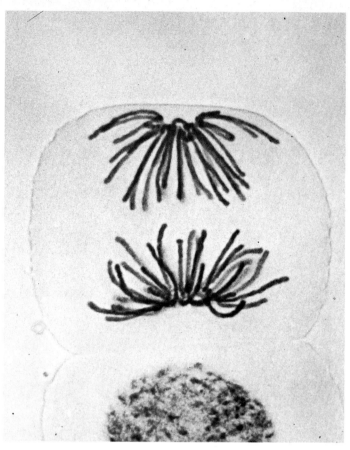

Plate 3*c*.   ANAPHASE
The chromatids have completely separated. The spindle is not visible in this photograph

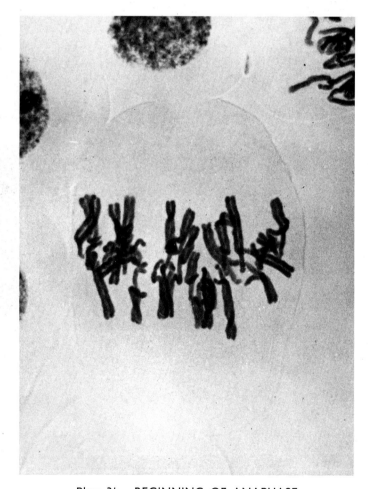

Plate 3*b*.   BEGINNING OF ANAPHASE
Each chromosome is seen to consist of two chromatids which are attached to the equator of the spindle and are beginning to separate

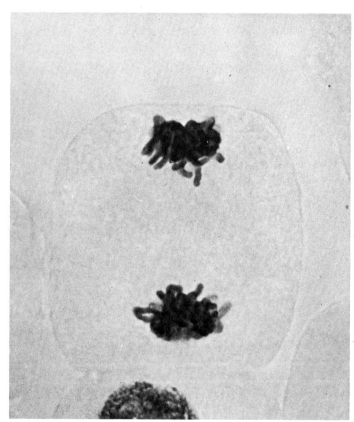

Plate 3*d*.   TELOPHASE
The chromosomes are becoming less distinct

Plate 3e. THE END OF CELL DIVISION
The daughter nuclei are separated by a new cell wall

Plate 3e. THE END OF CELL DIVISION
The daughter nuclei are separated by a new cell wall
(Plates I and 3a to e show photomicrographs of cells from the root tip of Lilium regale, from McLeish and Snoad, Looking at Chromosomes, Macmillan, 1958)

formed nevertheless.) By this time it is apparent that each chromosome consists of two parallel strands, called *chromatids* (Plate 3*b*), joined in one particular region, the *centromere*. In forming two chromatids, the chromosome *replicates*, that is, it produces an exact copy of itself but the two identical chromosomes remain in contact along their length. This replication occurs before prophase but is more evident during metaphase.

3. **Anaphase.** The two chromatids now separate at the centromere and begin to migrate in opposite directions towards either end of the spindle (Plate 3*c*). Experiments show that the spindle fibres play some part in separating the chromatids. The appearance is that of the chromatids first repelling each other at the centromere and then being pulled entirely apart by the shortening spindle fibres, although such a mechanism has not yet been verified.

4. **Telophase.** The chromatids, now chromosomes, collect together at the opposite ends of the spindle (Plate 3*d*) and become less distinct, probably by becoming uncoiled and therefore thinner. The one or more nucleoli reappear, and a nuclear membrane forms round each group of daughter chromosomes so that there are now two nuclei present in the cell. At this point in animal cells, the cytoplasm between the two nuclei constricts, and two cells are formed. Both may retain the ability to divide, or one or both may become specialized and lose their reproductive capacity. In plant cells, the cytoplasm does not constrict to form two new cells; instead, a new cell wall is formed across the cell in the region originally occupied by the equatorial plane of the spindle (Fig. 1.2*b* and Plate 3*e*).

Chromosomes

Chromosomes are so called because they take up certain basic stains very readily (*chromos*=colour, *soma*=body), but they can be observed by phase contrast microscopy in the unstained nuclei of dividing, living cells. When the cell is not dividing, the chromosomes cannot be seen in the nucleus, even after staining. Nevertheless, it is thought that they persist as fine, invisible threads, isolated patches of which will respond to dyes and show up as flakes or granules of deeply staining material. The chromosomes consist of protein and a substance called *deoxyribonucleic acid* (DNA: *see* p. 17), but the exact relationship between these two components in forming the chromosome is not known.

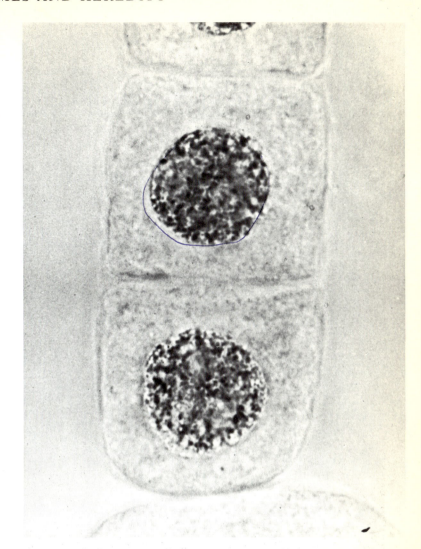

(World Health)

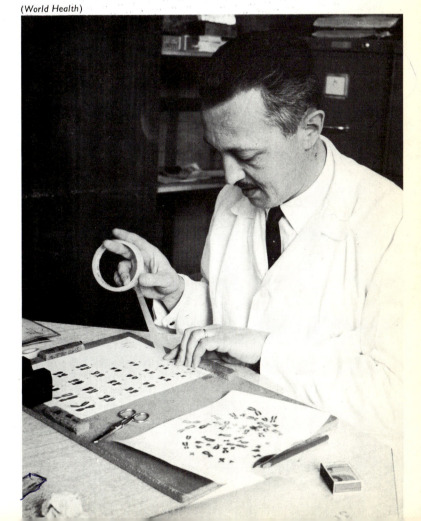

Plate 4. STUDYING HUMAN CHROMOSOMES
A member of a team of scientists under Professor Jerôme Lejeune at the Institut de Progénèse, Paris, identifies and prepares pictures of human chromosomes

Counts of chromosomes show that there is a definite number in each cell of any one species of plant or animal, e.g. mouse 40, crayfish 200, rye 14, fruit fly (*Drosophila*) 8, and man 46 (*see* also Fig. 1.5). It can also be seen (Plate 5) that the chromosomes exist in pairs, although not actually joined together, each pair having a characteristic length and, during anaphase, a characteristic shape (Plate 3*c*) governed by the position of the centromere at which the chromatids are pulled apart; e.g. a V shape if the centromere is central, or a √ shape if it is close to one end. In other words, human cell nuclei contain 23 pairs of chromosomes, mouse cells 20 pairs and so on, one member of each pair having been derived from the male and one from the female parent. The members of each pair are called *homologous chromosomes*, and the total number of chromosomes in each cell is called the *diploid number*.

From a consideration of the events at mitosis, it is apparent that each chromosome reproduces itself exactly when it forms two chromatids, which themselves become chromosomes. It is also seen that, after mitosis, the daughter nuclei will contain the same number of chromosomes as the parent cell nucleus.

Although the constituent chemicals of the cytoplasm of a cell are constantly being broken down and rebuilt from fresh material, the chemicals of the chromosomes remain remarkably stable. Other investigations show that during cell division, no protoplasmic material is shared so exactly as that of the chromosomes in the nucleus. Such evidence points to the chromosomes as the main source of the chemical information which determines that a cell should become like its parent cell, and that in their development, the cells of the organism will endow the animal or plant with all the characteristics of its species. Although the supporting evidence is very sketchily outlined here, the hereditary material must almost certainly lie on the chromosomes of the nucleus and be passed on to each daughter cell by the process of mitosis. In a similar way the nuclei of the gametes carry a set of chromosomes from the male and female parent, and these chromosomes determine that the zygote grows and develops to an animal or plant of the same species as the parents, reproducing to quite a minute degree, individual characteristics of both parents, e.g. flower colour or blood group.

(World Health)

Plate 5. PREPARING A 'KARYOGRAM'
The chromosome silhouettes from the photomicrograph on the left are cut out and arranged in order on the right-hand chart

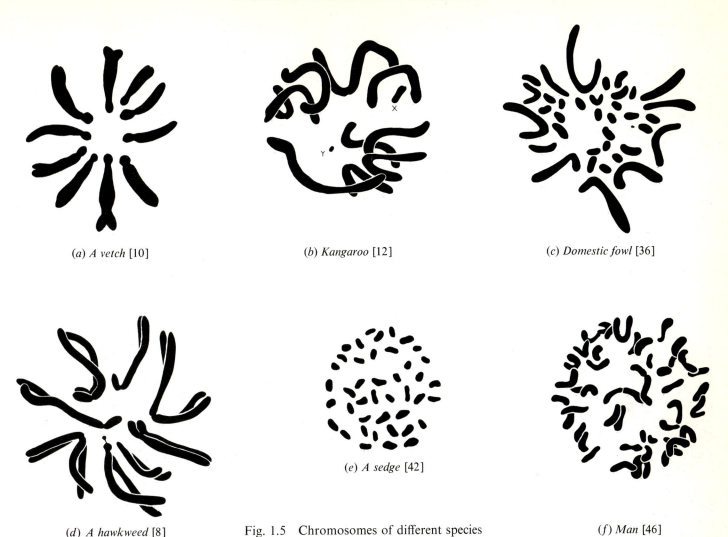

(a) *A vetch* [10]   (b) *Kangaroo* [12]   (c) *Domestic fowl* [36]

(e) *A sedge* [42]

(d) *A hawkweed* [8]   Fig. 1.5   Chromosomes of different species   (f) *Man* [46]

(From C. C. Hurst, *The Mechanism of Creative Evolution*, Cambridge University Press, 1933)

Genes. A gene is a theoretical unit of inheritance, theoretical in the sense that the word was coined long before chromosome structure was investigated in detail or the DNA theory of inheritance put forward. Today the gene is thought to consist of a group of chemicals situated on the chromosome, and the evidence for this is outlined below.

Chromomeres. During the early prophase of MEIOSIS (*see* p. 11), the chromosomes appear to consist of deeply staining granules connected by material which does not take up the stain (Fig. 1.6). The granules are called *chromomeres* and may represent regions where the chromosome is thrown into coils (Fig. 1.7). There is evidence to show that the chromomeres correspond to areas of gene activity on the chromosome, but little to indicate that the chromomeres are the genes themselves.

Fig. 1.7   Chromomeres may be the result of local coiling of chromosomes

Fig. 1.6   Homologous human chromosomes showing chromomeres

(From Yerganian, *American Journal of Human Genetics*, 1957, 1, **9**)

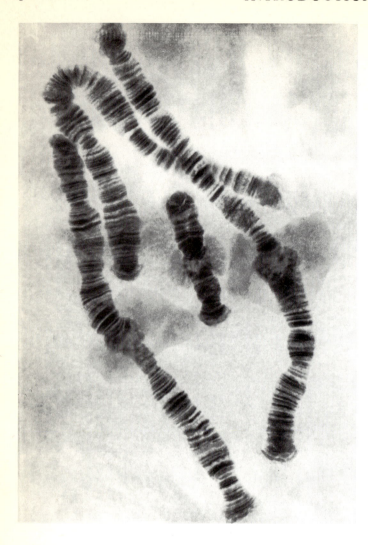

Giant chromosomes. Certain cells, e.g. those in the salivary glands of some fly larvae such as Drosophila (Fig. 1.8), contain unusually large chromosomes which are very useful material for the study of chromosome structure. These giant chromosomes are often a hundred times thicker and ten times longer than those in more typical cells and so can be studied in some detail with the light microscope which enables pronounced transverse bands to be seen (Plate 6 and Fig. 1.9). The explanation for these bands and the extraordinary size of the chromosomes is that the homologous chromosomes of each pair have remained together. Furthermore, they have replicated hundreds of times without the resultant chromatids separating or the nucleus dividing.

The compound chromosomes formed in this way are in a state of permanent prophase, so to speak, and it is probably because the chromomeres of adjacent chromatids match exactly that the chromosomes have a banded appearance (Fig. 1.10). Unlike most chromosomes in the later stages of prophase, the giant chromosomes do not appear to shorten by coiling, and this would account for their greater length.

Experimental work with fruit flies possessing these giant chromosomes has shown that when, perhaps through a genetical accident, bands are missing from chromosomes, certain characteristics fail to develop in the flies, and in many cases the bands have been identified with one or more genes (Fig. 1.11). So far, no genes have been located in a region of the chromosome having no bands, but on the other hand banded regions often seem to be the site of more than one gene.

Plate 6.   GIANT CHROMOSOMES

Four giant chromosomes from a cell in the salivary gland of the midge larva, *Chironomus tentans*, showing transverse banding. Magnification approx. × 350

(Courtesy of Professor Wolfgang Beermann, Max Planck Institute, Tübingen, from Sci. Amer., April 1964)

Fig. 1.8 (*b*) Notch deficiency in Drosophila

When a certain band is missing from the giant chromosome (Fig. 1.10), the wing of the fly exhibits this characteristic abnormality.

Fig. 1.8 (*c*) Vestigial wing

This abnormally small (and ineffective) wing occurs as a result of inheriting a single gene.

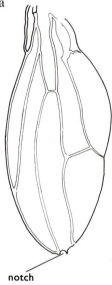

notch

Fig. 1.8 (*a*) Drosophila: the small fly used in many genetical experiments

eye

scutellum

bristle

wing

wing vein

examples of bands associated with genes

Fig. 1.9   Part of a giant chromosome from the salivary gland of a Drosophila larva

(From John A. Moore, *Heredity and Development*, Oxford University Press (New York), 1963. After C. B. Bridges, *Cytologia*, 1937)

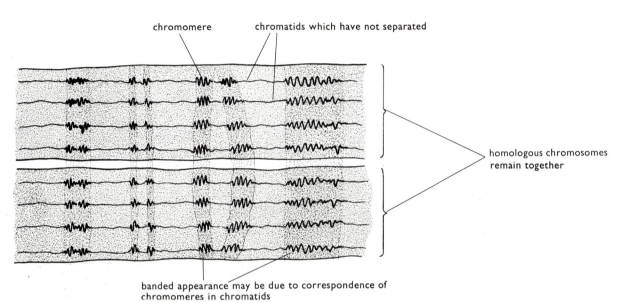

chromomere   chromatids which have not separated

homologous chromosomes remain together

banded appearance may be due to correspondence of chromomeres in chromatids

Fig. 1.10   Possible cause of bands in giant chromosomes

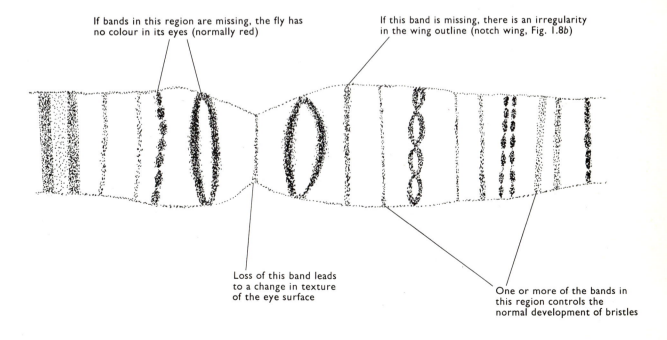

If bands in this region are missing, the fly has no colour in its eyes (normally red)

If this band is missing, there is an irregularity in the wing outline (notch wing, Fig. 1.8b)

Loss of this band leads to a change in texture of the eye surface

One or more of the bands in this region controls the normal development of bristles

Fig. 1.11   Part of salivary gland chromosome of Drosophila showing location of four genes

(From Curt Stern, *Principles of Human Genetics*, 2nd edn., W. H. Freeman & Company, 1960. After Slizynska, *Genetics*, 1938, 23)

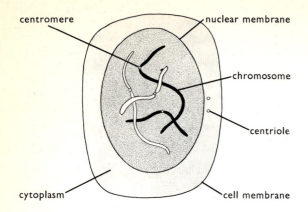

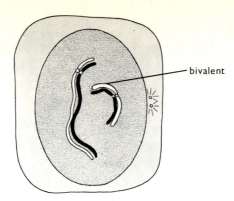

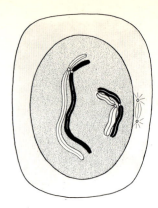

Prophase
(*a*) The diploid number of chromosomes appear

(*b*) Homologous chromosomes pair with each other, shorten and thicken

(*c*) Replication has occurred and the chromatids become visible

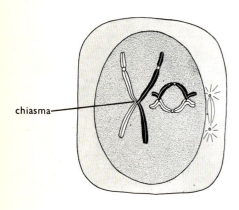

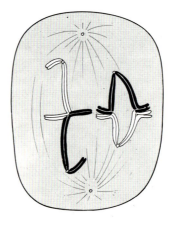

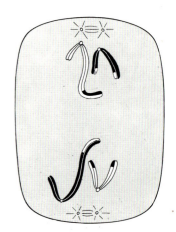

(*d*) Homologous chromosomes move apart except at the chiasmata where the chromatids have exchanged portions

Anaphase
(*e*) A spindle forms and homologous chromosomes move to opposite ends taking the exchanged portions with them

(*f*) Homologous chromosomes separated but not enclosed in nuclear membranes

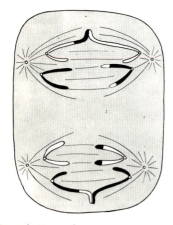

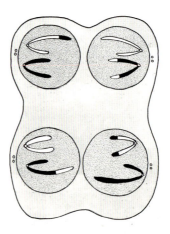

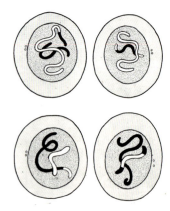

Second meiotic division
(*g*) Spindles form at right angles to the first one and the chromatids separate

Telophase
(*h*) Four nuclei appear, each enclosing the haploid number of chromosomes

(*i*) Cytoplasm divides to form four gametes

Fig. 1.12   Meiosis in a gamete-forming cell

Gene function

The picture that emerges from this and other evidence is that the genes which determine the characteristics of the organism are somehow arranged in line down the chromosome. These genes control the production of enzymes which in turn determine what functions go on in a cell, and eventually in the organs and entire organism. If anything happens to a gene it will affect the organism, e.g. in mice there is a gene which determines that the coat will be coloured. If this gene is missing, the mouse will be without pigment; it will be white with pink eyes. In this case, as in many others, more than one gene will in fact play a part in determining the characteristic.

The number of genes in man is not known but it could be about 1000 per chromosome. At mitosis, each chromomere, and therefore each of any of the genes it comprises, is exactly reproduced.

Two problems arise from this account. Since every cell in the body carries an identical set of chromosomes and since cell structure and function are determined by the genes on the chromosomes, why is not every cell of the body identical? Furthermore, what possible part can a gene for brown eyes play when it is in the nucleus of a cell lining the stomach wall? Briefly, when we follow the development of a particular cell, it seems that the way in which one of its genes will affect the cell depends not only on the gene itself but also on the physiology of the cell, which in turn is related to its particular position in the body. For example, the chemical environment in a certain cell in the scalp allows the gene for black hair to operate in a particular way. Just what the same gene does in another part of the body is not certain, but it is known that most genes have more than one effect, and the characteristic by which they are recognized is not necessarily their most important function; e.g. the genes responsible for producing colour in the scales of one kind of onion also confer a resistance to fungus disease because they determine the presence of certain chemicals which act as a fungicide. The colour, however, is the more obvious characteristic. The gene in Drosophila which produces the effect of diminutive wings also reduces the expectation of life to half that of normal flies. The wing characteristic is the more immediately obvious effect but the effect on life span may be far more important and damaging to the species. The idea that the expression of a gene depends to some extent on the physiology of the cell and the situation in which it finds itself is illustrated by the experimental work with certain amphibian embryos. If a piece of tissue, which would normally become skin, is taken from the abdomen and grafted into a region overlying the developing eye, the graft will be incorporated into the eye as a lens. It has the same chromosomes and genes, but its new position has altered its fate. This effect is by no means true of all animals and is certainly not the case in insects in which the fate of individual cells seems to be determined at a very early stage in development and is not affected by moving the cells to a new situation.

Meiosis (Fig. 1.12)

Cells in the reproductive organs, which are going to form gametes, e.g. sperms and ova, undergo a series of mitotic divisions resulting, in the case of male gametes, in a vast increase of numbers. The final divisions, however, which give rise to mature gametes are not mitotic. Instead of producing cells with 46 chromosomes in man, they form gametes with only 23 chromosomes. When, at fertilization, there is a fusion of the two gametes, the resulting zygote contains the diploid number of 46 chromosomes, and this number is present in all the cells of the offspring. The halving of the chromosome number which occurs at gamete formation, ultimately maintains the diploid number of chromosomes characteristic of the species. If gametes were produced by mitosis, a human egg and sperm would each contain 46 chromosomes and when they fused at fertilization would give rise to a zygote with 92 chromosomes. The gametes from the resulting organism would in turn give rise to offspring with 184 chromosomes and so on.

1. **Prophase.** In meiosis, the chromosomes appear in the nucleus (i.e. they become visible when the cell is fixed and stained or observed by phase contrast microscopy) in much the same way as described for mitosis, but although it is reasonably certain that two chromatids are present in each chromosome, the chromosomes still appear to be single threads. Other differences from mitosis seen at this stage are the appearance of chromomeres and the failure of the chromosomes to shorten by coiling.

In complete contrast to mitosis, the homologous chromosomes now appear to *attract* each other and come to lie alongside so that the centromeres and chromomeres correspond exactly. The pairs of chromosomes so formed are called *bivalents*, e.g. a cell with a normal complement of 6 chromosomes would have, at this stage, 3 bivalents.

In this paired state the chromosomes shorten and thicken by coiling, and now each chromosome is seen to consist of two chromatids. As soon as this occurs, however, the pairs of chromatids seem to repel each other and move apart, except at certain regions called *chiasmata*. In these regions the chromatids appear to have broken and joined again but to a different chromatid. The significance of this exchange of sections of chromosomes or 'crossing over', is discussed on p. 14. All these changes occur during prophase while the nuclear membrane is still intact.

2. **Metaphase.** The nuclear membrane disappears, a spindle is formed and the bivalents approach the equatorial region.

3. **Anaphase.** The paired chromatids of each bivalent now continue the separation that began in prophase and move to opposite ends of the spindle in a manner superficially similar to that of chromatids in mitosis. The outcome is that only half the total number of paired chromatids reaches either end of the spindle. Thus although there may originally have been 6 chromosomes in the nucleus, there are now only 3 paired chromatids at either end of the spindle.

4. **Second meiotic division.** A nuclear membrane does not usually form round the paired chromatids at this stage. Instead, they immediately undergo a second division at right angles to the first, only this time the chromatids are separated.

5. **Telophase.** The four groups of chromatids are now enclosed in nuclear membranes so forming four nuclei, each containing half the diploid number of chromosomes (the *haploid* number). Finally, the cytoplasm divides to separate the nuclei, giving rise, in the case of males at least, to four gametes (Plate 2). In sperm formation (*spermatogenesis*) in most animals the four cells will develop 'tails' to become sperms.

Formation of ova: oogenesis. During the formation of ova, the cytoplasm is not shared equally. After the first meiotic division, one of the daughter nuclei receives the bulk of the cytoplasm and the other nucleus is separated off with only a vestige of cytoplasm to form the first *polar body*, which, although it may undergo the next stage of its meiotic division, cannot function as an ovum and subsequently degenerates.

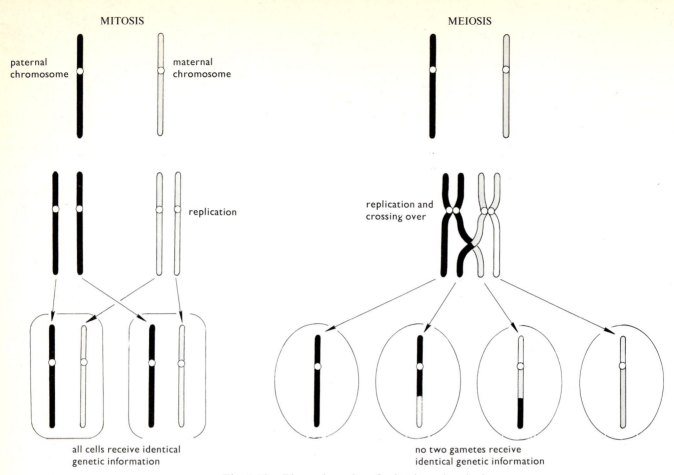

Fig. 1.13 The end results of mitosis and meiosis

In a similar way, the next meiotic division of the remaining egg nucleus produces a second polar body and a mature ovum. In many vertebrates, the first polar body is not formed until after the potential ovum is released from the ovary, and the second polar body, until after the penetration of the sperm in fertilization. In man, when a sperm bearing 23 chromosomes fuses with a 23-chromosome ovum, a 46-chromosome zygote is formed.

New combinations of genes in the gametes. In mitosis the full complement of chromosomes derived from both parents is first doubled and then shared equally between daughter cells. The result of this replication is that the daughter cells receive identical genetic information. In meiosis, on the other hand, the genetic information on the chromosomes is not shared in exactly the same way between all the gametes (Fig. 1.13). If homologous chromosomes were identical in their gene content, this variability of chromosome distribution would have no effect. Since, however, an individual's parents are likely to be genetically dissimilar in many respects, there will be many gametes with combinations of genes quite different from either of the individual's parents.

Fertilization (Fig. 1.14)

The cytoplasm of the sperm fuses with that of the ovum and the male nucleus passes into the ovum, coming to lie alongside the egg nucleus: the zygote is formed, but there is no fusion of nuclear material at this stage. Each nucleus simultaneously undergoes a mitosis with the axes of the spindles parallel to each other, but at the telophase stage the adjacent chromatids, originally from different parents, become enclosed in the same nuclear membrane, thus restoring the diploid number of chromosomes. These events are followed by the first cleavage, the zygote dividing into two cells. Subsequent mitotic division produces a multicellular organism with the diploid number of chromosomes in all its cells.

Recombination of genes in the zygote. In man there are about 200 million sperms in a single ejaculation and about 300 to 400 eggs produced in a reproductive lifetime. The eggs will have a genetic content differing from the sperms (e.g. wife with curly red hair; husband with straight black hair). When the chromosomes of the sperm and egg combine, the zygote may contain genes for red, black, curly and straight hair. The genes for curly and black are dominant (p. 23) to those for straight and red, so that although the child may carry all 4 genes, his hair will be curly and black—a new combination of characters not represented in either parent. When this child grows up, his own gametes are formed. If the genes for hair colour and curliness are on different chromosomes, he may produce sperms with 4 alternative combinations of these genes; namely, red and straight, black and straight, red and curly, black and curly.

Determination of sex. In humans, one pair of the smallest chromosomes is known to determine sex. In the female, these two chromosomes are entirely homologous and are called the *X chromosomes*, while in the male, one is smaller and is called the *Y chromosome* (Fig. 1.5b). Femaleness normally results from the possession of two X chromosomes and maleness from possession of an X and a Y chromosome. At meiosis, the sex chromosomes are separated in the same way as the others (Fig. 1.15), so that all the female gametes will contain an X chromosome, but half the male gametes will contain an X and

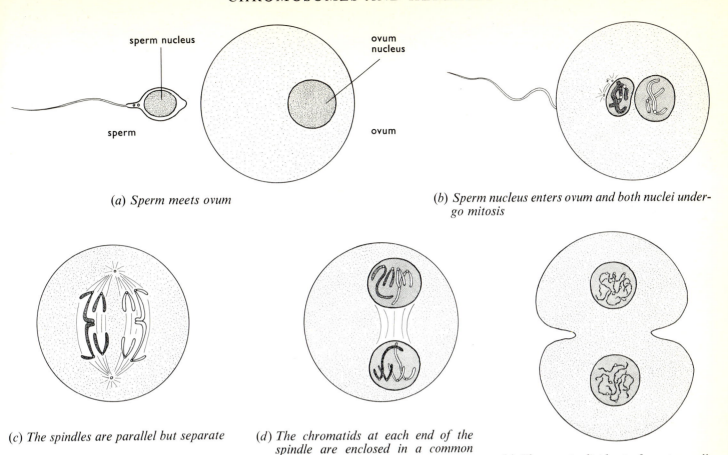

(a) Sperm meets ovum

(b) Sperm nucleus enters ovum and both nuclei undergo mitosis

(c) The spindles are parallel but separate

(d) The chromatids at each end of the spindle are enclosed in a common nuclear membrane

(e) The zygote divides to form two cells

Fig. 1.14   Fertilization (polar bodies not shown)

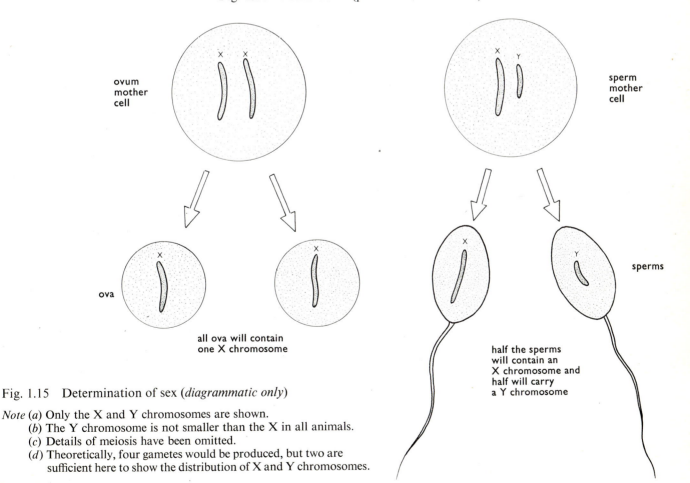

Fig. 1.15   Determination of sex (diagrammatic only)

Note (a) Only the X and Y chromosomes are shown.
   (b) The Y chromosome is not smaller than the X in all animals.
   (c) Details of meiosis have been omitted.
   (d) Theoretically, four gametes would be produced, but two are
       sufficient here to show the distribution of X and Y chromosomes.

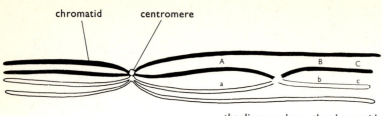

the diagram shows the chromatids breaking at the chiasma but it is not known if this is what actually occurs

(a) PROPHASE

Homologous chromosomes have paired up

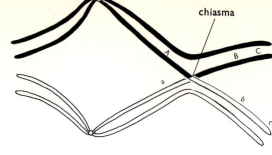

(c) METAPHASE

The homologous chromosomes seem to repel each other except at the chiasma

the terminal portions of the adjacent chromatids have become attached to the opposite chromatid

(b) PROPHASE

Fig. 1.16   Crossing over

half will contain a Y chromosome. If a Y-bearing sperm fertilizes an ovum, the zygote will be XY and give rise to a boy. Fertilization of an ovum by an X-bearing sperm gives an XX zygote which develops to a girl. There should be an equal chance of X or Y sperms meeting an ovum, and therefore equal numbers of boy and girl babies should be born. In fact, slightly more boys than girls are born in most parts of the world. The reason for this is not clear, but it also happens that the mortality rate for boy babies and men is slightly higher than for girl babies and women, which tends to restore the balance.

Although the X and Y chromosomes determine sex, it does not necessarily follow that male and female characteristics are determined by genes found only on the sex chromosomes. In man, genes for male and female characters may be scattered fairly evenly throughout all the chromosomes, but the presence of the Y chromosome in an XY zygote may tip the balance in favour of maleness. Femaleness results from the absence of the Y chromosome but this is not the case in all the animals studied; e.g. it is true for the mouse but not for Drosophila.

Linkage and crossing over. During that early stage of meiosis when homologous chromosomes pair up, the chromatids of maternal and paternal chromosomes may exchange portions as mentioned on pp. 11 and 12 and shown in Fig. 1.16. In the absence of crossing over, the maternal genes A-B-C on the same chromosome would always appear together no matter how the chromosomes were assorted in meiosis. Similarly the paternal genes a-b-c would always remain together. For example, in Drosophila, since the genes for black body, purple eyes and vestigial wings occur on the same chromosome, one might expect that a black-bodied Drosophila would always have purple eyes and vestigial wings. Crossing over between chromatids, however, gives the possibility of breaking these *linkage groups*, as they are called, so that new combinations, ABc, Abc, aBC, abC, aBc, AbC, could arise in the gametes,

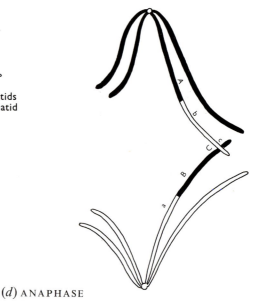

(d) ANAPHASE

The chromosomes separate, but as a result of crossing over, the genes A, B, C and a, b, c on the 'inner' chromatids are rearranged

two of them being black body with red eyes and normal wings, or black body with purple eyes and normal wings.

Genes at opposite ends of the chromosome are likely to be separated more frequently by crossing over than are closely adjacent genes (Fig. 1.17). The frequency with which certain characteristics appear together in a plant or animal enables the geneticist to judge the extent to which the genes are linked. If in a certain kind of plant, red petals always appeared on tall plants and white petals on short ones, it could be assumed that the genes for red and tall were on the same chromosome and very close together. If these two characteristics appeared together in 80 per cent of all the plants, it would suggest that, although the two genes were on the same chromosome, some crossing over had taken place between them, allowing tallness and white petals or shortness and red petals to appear in 20 per cent of the progeny.

In this way, 'maps' have been made showing the degree of linkage between genes. The number and, in some cases, the relative length of these linkage groups corresponds to the number and relative length of the actual chromosomes (Fig. 1.18)—further evidence that chromosomes carry the genes in a linear sequence.

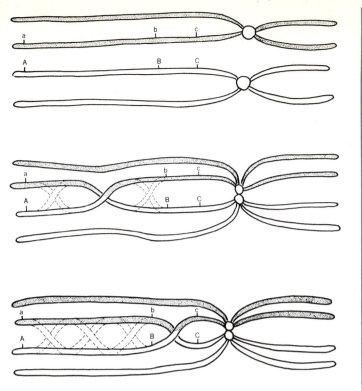

Fig. 1.17 Frequency of crossing over between genes on homologous chromosomes. (*Crossing over is likely to occur more frequently between A and B than between B and C if the chiasmata are randomly distributed*)

(From Professor K. Mather, *Genetics for Schools*, Modern Science Memoir No. 31, John Murray, 1953)

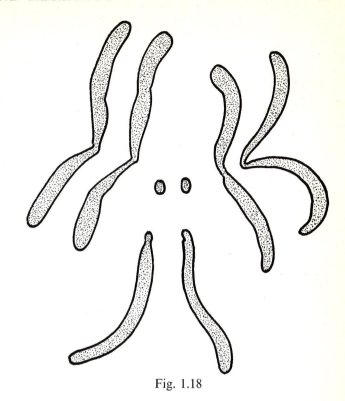

Fig. 1.18

(*a*) THE CHROMOSOMES OF DROSOPHILA AT MITOTIC METAPHASE

(*b*) LINKAGE GROUPS IN DROSOPHILA (*see* Fig. 1.8, p. 8)

I. y = yellow (body)
w = white (eye)
cv = crossveinless (wings)
ct = cut (wings)
v = vermilion (eyes)
s = sable (body)
f = forked (bristles)
bb = bobbed (bristles)

II. al = aristaless (aristae)
dp = dumpy (wings)
d = dachs (legs)
b = black (body)
pr = purple (eye)
vg = vestigial (wings)
c = curved (wings)
px = plexus (wing veins)
sp = speck (wing base)

III. ru = roughoid (eye)
h = hairless (antennae)
th = thread (antennae)
p = pink (eye)
sr = stripe (body)
e = ebony (body)
ro = rough (eye)
ca = claret (eye)

IV. ci = cubitus interruptus (wing veins)
ey = eyeless (eye)
sv = shaven (bristles)
gvl = grooveless (scutellum)

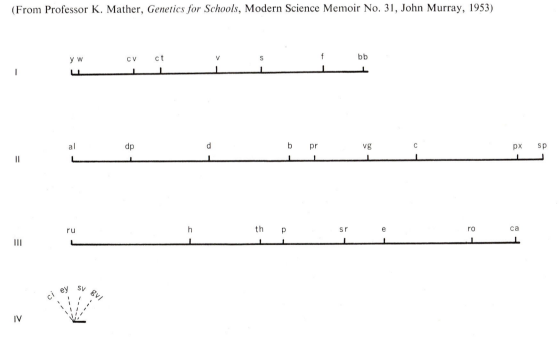

The relative positions of the genes are worked out from the frequency of crossing over. Note that there are 4 linkage groups and 4 pairs of chromosomes; 2 long, 1 medium and 1 very short pair of chromosomes, which corresponds to the relative length of the linkage groups.
Only a few of the known genes are indicated in this drawing.

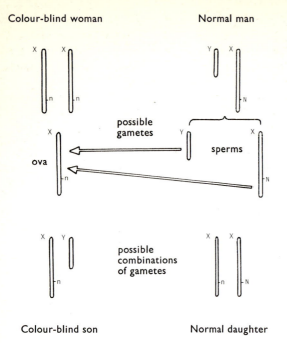

Colour-blind woman   Normal man

possible
gametes

ova   sperms

possible
combinations
of gametes

Colour-blind son   Normal daughter

N = normal colour vision

n = colour blindness

Fig. 1.19   SEX LINKAGE, showing the possible distribution of X and Y chromosomes between the gametes and the chances of combination in the zygotes

The theoretical expectations are that all the girls will be normal but half of them will be carriers, half the boys will be normal and half of them colour blind. The types of children expected from the marriage between a woman carrier and a colour-blind man, or a normal woman and a colour-blind man can be worked out in a similar way.

Other X-linked factors are *haemophilia* and brown enamel on the teeth. Haemophilia causes a delay in the clotting time of the blood. Although there are two kinds of sex-linked haemophilia, at least three other clotting disorders are known which are controlled by genes not on the sex chromosomes.

Sexual characteristics such as bass voice, beard and muscular physique in males, mammary glands and wide pelvis in females, are not the result of sex-linked genes but the different expressions of the same genes present in both sexes. Both sexes carry genes controlling the growth of hair, mammary glands and penis but in the physiological environment of maleness or femaleness they have different effects, with the result that e.g. the mammary glands in males are small and functionless; the penis in females is represented by only a small organ, the *clitoris*.

Only a few rare abnormalities are thought to be linked to the Y-chromosome and even these are now open to doubt.

Sex linkage. Certain genes which occur on the X chromosome are more likely to affect a male than a female. The gene or genes for a certain form of colour blindness in man are carried on the X chromosome. Normal vision is dominant (p. 23) to colour blindness so that if a colour-blind woman, who must be homozygous (p. 23) for the character, marries a normal man, all their sons but none of their daughters will be colour blind. This can be explained by the fact that the Y chromosome is homologous with only a small section of the X chromosome and the non-homologous part of the X chromosome carries genes which are not represented on Y. It is assumed that the Y chromosome plays no part in the determination of colour vision. Fig. 1.19 shows how this type of sex linkage produces its effect.

The normal daughter is heterozygous (*see* p. 23) for colour blindness and is therefore a 'carrier' for the recessive gene. If she marries a normal man the possible combinations of genes in the children are shown by:

Parents:	XN Xn		XN Y	
	carrier woman		normal man	
Gametes:	XN	Xn	XN	Y
Possible combinations of gametes:	XN XN	Xn XN	XN Y	Xn Y
	normal girl	girl carrier	normal boy	colour-blind boy

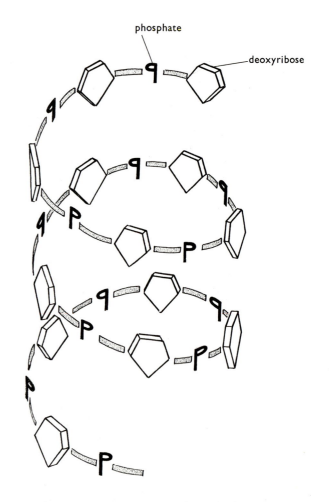

phosphate

deoxyribose

Fig. 1.20   The sugar-phosphate chain in DNA

The diagram shows part of the molecule which forms the backbone of DNA—a coiled chain of deoxyribose and a 5-carbon sugar alternating with phosphate groups

How genes work

It was mentioned earlier (p. 5) that chromosomes consist of protein and a nucleic acid, deoxyribonucleic acid (DNA). Although the precise relationship between the DNA and the protein is not known, the structure of DNA has been intensively studied. This chemical consists of long molecules coiled in a double helix. The strands of the helix are chains of sugars and phosphates, the sugar being a 5-carbon compound, *deoxyribose* (Fig. 1.20). The two helices in a DNA strand are linked together by cross-bridges made by pairs of organic nitrogenous bases joined to the sugar molecules (Figs. 1.21 and 1.22). Although there are only four principal kinds of base in the DNA molecule, *adenine, cytosine, thymine* and *guanine*, it is thought that the sequence of these bases is the important factor in heredity, and that a gene may consist of a particular sequence of up to 1000 base pairs in a DNA molecule.

The different sequences of bases *along the length* of the DNA molecule seem to act like a code, instructing the cell to make certain proteins, the order of bases indicating the sequence of amino acids to be joined up in order to make the protein. For example, the sequence CAA (cytosine-adenine-adenine) specifies the amino acid *valine*: three thymines in a row, TTT, specify *lysine*, while AAT specifies *leucine*. So the sequence of bases CAA-TTT-AAT would direct the cell to link up the amino acids valine-lysine-leucine to make the appropriate peptide, and in a similar way proteins are formed.

Most of the proteins made are enzymes which direct the pattern of chemical activity in the cell. Thus DNA, by determining the kinds of enzyme formed in the cell, will control the cell's activities. This in turn will affect the nature of the cell, the organ of which it is a part and eventually the organism as a whole. A change in the sequence of bases in the DNA molecule will result in a different order of amino acids and, hence, a different, and probably ineffective, enzyme. This will usually act adversely on the metabolism of the cell.

A rabbit with coloured fur has a gene which controls the production of the enzyme *tyrosinase*. This enzyme converts *tyrosine*, a colourless amino acid, to *melanin*, a black pigment. An albino rabbit has no gene for tyrosinase production, and consequently no pigment is formed from the tyrosine in its body.

Normal humans have a gene which controls the production in the blood of an enzyme which accelerates the breakdown of a chemical, *alcapton*. Persons having no gene for the enzyme excrete in the urine unchanged alcapton which darkens on exposure to the air. This relatively harmless effect is associated with pigmentation in other parts of the body and, later, with arthritis. The condition is inherited as a recessive factor (p. 23). This is a rather peculiar example of the mechanism of

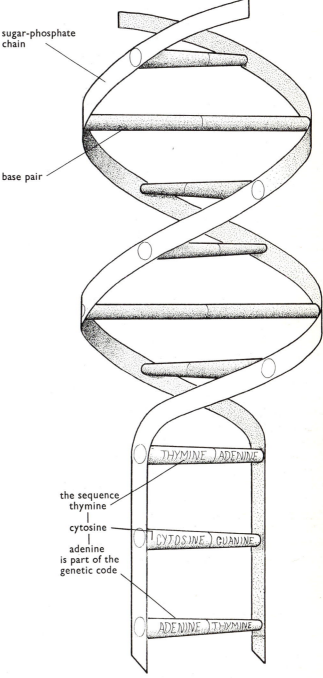

sugar-phosphate chain

base pair

the sequence
thymine
|
cytosine
|
adenine
is part of the
genetic code

THYMINE   ADENINE

CYTOSINE   GUANINE

ADENINE   THYMINE

Fig. 1.21   Part of a DNA molecule
($\bigcirc$ =*deoxyribose*)

Fig. 1.22   The drawing shows schematically, part of a DNA molecule. The lower part is shown uncoiled to emphasize the position of the base pairs

inheritance, but if a gene controlled the production of an enzyme essential in a much earlier stage of a series of vital reactions, the absence of the gene could have devastating effects even to the extent of causing premature death. Conversely, since normal physiology is the result of hundreds of chemical changes catalyzed by hundreds of enzymes, it is not surprising that characteristics such as intelligence, stature and activity come under the influence of many genes.

The role of RNA. Cell proteins are not made by the nucleus but by the ribosomes in the cytoplasm. The coded information in the nuclear DNA must somehow be transposed to the cytoplasm. This transfer is carried out by a nucleic acid, *ribonucleic acid* (RNA), which differs slightly in composition from DNA. Each part of the DNA that is active in the chromosomes builds up a replica of itself in RNA. These lengths of RNA, called *messenger RNA,* become detached from the chromosomes,

leave the nucleus through the nuclear pores and become attached to ribosomes (Fig. 1.23a and b).

In the cytoplasm there are many amino acids and some of these become attached to a second kind of RNA called *transfer RNA*. Three of the bases on each molecule of transfer RNA correspond to one of the groups of three bases on the messenger RNA. The amino acid molecules attached to the transfer RNA will eventually encounter a ribosome to which is attached a strand of messenger RNA. If the section of messenger RNA attached to the ribosome has the three bases that correspond to those on the transfer RNA, the latter will combine temporarily with the messenger RNA. As the messenger RNA moves along the ribosome, the triplets of bases will be presented in turn, and each triplet will allow its corresponding transfer RNA to attach itself, bringing an amino acid with it. The amino acids are linked together by chemical bonds in the correct sequence to make polypeptides and proteins (Fig. 1.23 d–f).

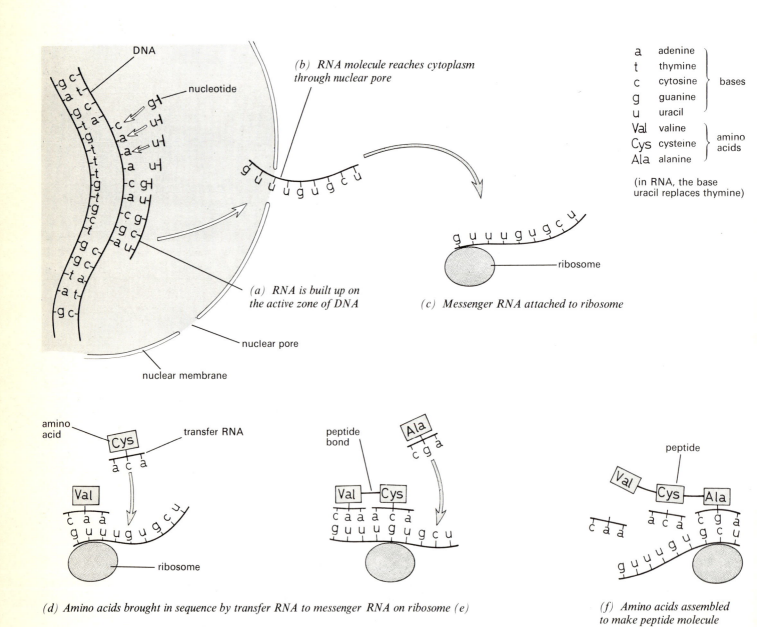

Fig. 1.23   The role of DNA and RNA

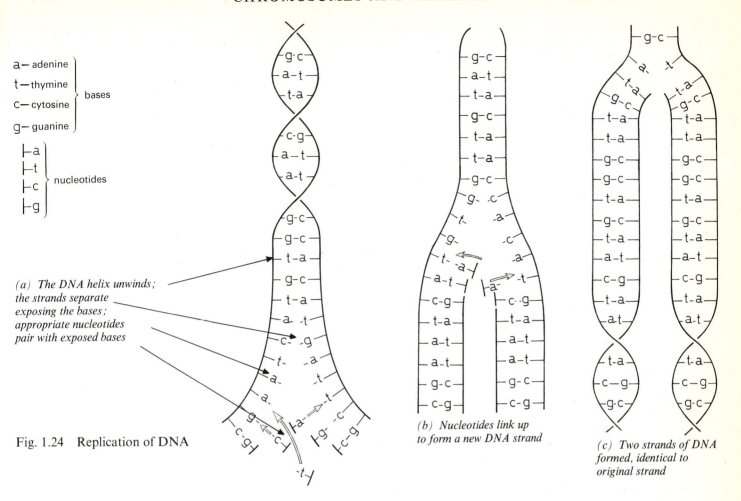

a — adenine
t — thymine } bases
c — cytosine
g — guanine

├a
├t } nucleotides
├c
├g

(a) The DNA helix unwinds; the strands separate exposing the bases; appropriate nucleotides pair with exposed bases

Fig. 1.24   Replication of DNA

(b) Nucleotides link up to form a new DNA strand

(c) Two strands of DNA formed, identical to original strand

Replication of genes. When a chromosome forms two chromatids prior to cell division, its constituent DNA also replicates. It is thought that the component strands of each double helix unwind, exposing the organic bases. In the nucleoplasm are the corresponding bases attached to sugar and phosphate molecules. This combination of organic base, sugar and phosphate is called a *nucleotide*. The bases on the appropriate nucleotides become attached to the exposed bases of the separated DNA strands, and so build up a new molecule of DNA with exactly the same sequence of bases as the original partner. In this way the genetic code is preserved intact to be passed on with the chromosomes to the new cell (Fig. 1.24).

Mutations

A mutation is a spontaneous change in a gene or a chromosome which may produce an alteration in the characteristic under its control. Fig. 3.6 (p. 37) shows a chromosome mutation. A fairly frequent form of mental deficiency known as Down's syndrome (mongolism) results from a chromosome mutation in which the ovum carries an extra chromosome, so that the child has 47 chromosomes in his cells instead of 46.

Gene mutations. During replication, nucleotides in the nucleoplasm pair with the exposed nucleotides on the 'unwound' DNA strand. Normally, the base, adenine, pairs with thymine while the base guanine pairs with cytosine. As a result of this, the new DNA strand is identical to that which previously occupied the same position and the genetic information on the double strand is unchanged. Sometimes, however, due to a temporary change in one of the bases at

replication, adenine pairs with cytosine and guanine pairs with thymine so that one of the new strands carries a different sequence of bases from its predecessor. This is a mutation. When this new strand replicates, the 'error' is perpetuated.

The RNA built up from the altered DNA strand will also be different and the triplet which includes the 'wrong' base will cause an amino acid different from usual to be built into the protein made by the RNA. Consequently, the properties of the protein will be altered. If the protein is an enzyme, the 'incorrect' amino acid may so alter its properties as to make it ineffective. In the absence of this enzyme, the cell's physiological activities may be drastically impaired. An example of such a gene mutation which affects the production of haemoglobin in the red blood cells is described on p. 37.

A mutation in a single cell may not be very important in heredity, but if the cell is a gamete mother cell, a gamete or a zygote, the entire organism arising from this cell may be affected. Since the mutant form of the gene is inherited in the usual way, the mutation may persist in subsequent generations.

On the whole, genes are stable structures because DNA is a stable chemical, but once in a hundred thousand replications or more a gene may mutate. The frequency with which certain genes mutate has been estimated for many abnormalties in man and in experimental animals. Most mutations that produce an observable effect seem to be harmful if not actually lethal. This is not surprising, since any change in a well—but delicately-balanced organism is likely to upset its physiology. Most mutations, however, are recessive and so may not find expression in the heterozygous form (p. 23). The variants of Drosophila already mentioned, e.g. vestigial wings, black body, purple eyes, are the result of single gene mutations from the normal, wild type.

In humans there occurs a form of dwarfism known as *achondroplastic dwarfism*, in which the limb bones do not grow normally. This condition arises as a result of a dominant mutation.

Mutation rate. One cannot predict when a gene is going to mutate but the frequency of its occurrence can be determined in some cases; for example in achondroplastic dwarfism it is possibly as high as one mutation in 20 000, compared with one in 100 000 for many genes. The rate of mutation is characteristic of particular genes in particular species, but the frequencies are such that in a human ejaculate of, say, 200 million sperms, there are likely to be a considerable number of nuclei bearing gene and chromosome mutations.

Exposure to radioactivity, X-rays and ultraviolet radiation is known to increase the rate of gene and chromosome mutation.

Radiation and mutations. The cause of mutation is not known, but exposure to X-radiation, gamma-radiation, ultra-violet light, etc., is known to cause an increase in the mutation rate in experimental animals such as fruit flies and mice. The artificially-induced mutations are the same as those which occur naturally, but the frequency with which they occur is greatly increased.

There is a fairly constant background of radiation on the Earth's surface as a result of cosmic rays. Individuals also receive radiation from X-rays used in medicine, television tubes and luminous watch dials. Workers in atomic power stations and other people handling radioactive materials in industry or research may receive additional radiation. The radioactive fall-out from atomic explosions has increased the background radiation.

It is of obvious importance to assess the effect of any increase in radiation on the health of individuals and, as a result of mutations in their reproductive cells, the health of their children.

There is, so far, insufficient information to determine the correlation between the radiation dose and the mutation rate in man. The maximum safe dose in respect of direct effects on the individual, e.g. leukaemia, is still a matter of controversy. Nevertheless it is probably safe to say that any increase in the mutation rate is likely to have harmful effects on the population. Consequently, the exposure of individuals to the hazards of radiation is limited, though somewhat arbitrarily, by law.

PRACTICAL WORK

1. *Squash preparation of chromosomes using acetic orcein.*

Material. *Allium cepa* (onion) root tips. Support onions over beakers or jars of water using tooth-picks as shown in Fig. 1.26. Keep the onions in darkness for several days until the roots growing into the water are 2–3 cm long. Cut off about 5 mm of the root tips, place them in a watch glass and

(*a*) cover them with 9 drops acetic orcein and 1 drop molar hydro-chloric acid;

(*b*) heat the watch glass gently over a very small Bunsen flame till steam rises from the stain, but do not boil;

(*c*) leave the watch glass covered for at least five minutes;

(*d*) place one of the root tips on a clean slide, cover with 45 per cent acetic (ethanoic) acid and cut away all but the terminal 1 mm;

(*e*) cover this root tip with a clean cover-slip and make a squash preparation as described below.

Making the squash preparation. Squash the softened, stained root tips by lightly tapping on the cover-slip with a pencil: hold the pencil vertically and let it slip through the fingers to strike the cover-slip. The root tip will spread out as a pink mass on the slide; the cells will separate and the nuclei, many of them with chromosomes in various stages of mitosis (because the root tip is a region of rapid cell division) can be seen under the high power of the microscope ($\times 400$).

To make a permanent preparation. The slide prepared as above will last for weeks if the cover-slip is 'ringed' with a suitable material, e.g. melted wax or nail varnish, to prevent evaporation of the liquid, but a more permanent slide can be made by using, at stage (*d*) above, a slide smeared with egg albumen. Place a small drop of glycerol albumen (*see* (iv) below) on the slide and then smear it over with the finger, as if trying to wipe it all off, thus leaving a very thin film. Pass the slide through a Bunsen flame until it begins to 'smoke' but not to the point of charring. Make the squash preparation as before and then invert the slide in a shallow dish of 10 per cent acetic acid (Fig. 1.25) until the cover-slip falls off, leaving the squashed root tip adhering to the slide. Dip the slide in 80 per cent alcohol and then into 'Ethex' (ethylene glycol monoethyl ether) to dehydrate it for 10–30 minutes.

Finally, mount the preparation by lowering a cover-slip over a drop of Euparal covering the root cells (Fig. 1.27).

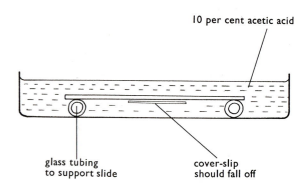

Fig. 1.25   Floating-off the cover-slip

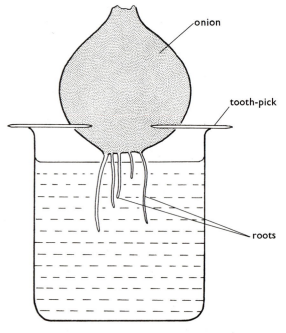

Fig. 1.26   Method of supporting an onion to promote growth of roots

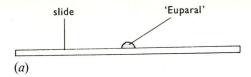

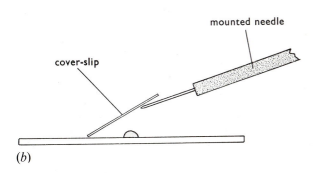

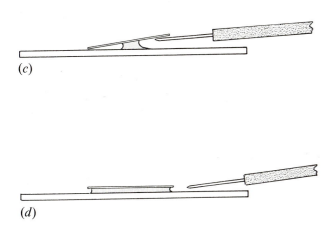

Fig. 1.27 Method of applying cover-slip in a permanent preparation

3. *Breeding experiments with Drosophila.*

(Chapter 2 should be studied before attempting these experiments.)

Drosophila is a small fly (Fig. 1.8a, p. 8) which is easy to breed in large numbers in the laboratory. By carrying out controlled cross-breeding experiments with mutant forms and wild types, it is possible to illustrate and investigate some of the principles of heredity.

Sources. Wild type and several mutant strains can be obtained from the usual biological supply firms.

Culture medium. Mix together 50 g maize meal, 15 g agar, 13 g dried yeast, 25 g brown sugar, 800 cm³ water and boil gently for several minutes. Dissolve 2 g 'Moldex' (a mould-inhibitor) in 40 cm³ boiling water and add it to the mixture. Pour the mixture into specimen tubes 100 × 25 mm to a depth of 20 mm, plug the tubes with cotton wool wrapped in butter muslin and sterilize in an autoclave at 1 kg/cm² for 15 min. On the day before introducing the flies, add 3 drops of a suspension of fresh yeast to each tube.

The tubes will hold about one hundred flies for experiments but for maintaining stocks of Drosophila, wide-mouthed bottles, e.g. half-pint milk or cream bottles should be used and new cultures started every five or six weeks (Fig. 1.28).

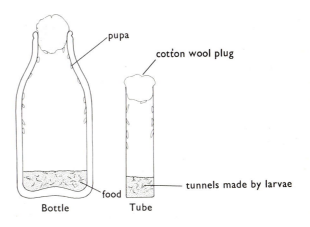

Fig. 1.28 Drosophila cultures

2. *Meiosis*

To see chromosomes in meiosis, the same technique is adopted using material where gamete mother cells are giving rise to gametes, e.g. the developing anthers of bluebell flowers, where pollen grains are being produced.

Dig the bulbs early in the year when the leaves just appear above the ground and dissect out the inflorescence, fixing them straight away in Clarke's fluid for 24 hours. Select flowers of a suitable age, i.e. those higher up on the stalk than the ones in which the anthers are beginning to turn yellow. Dissect out the anthers and stain them in acetic orcein as before.

Preparation of reagents

(i) **Clarke's fluid.** 25 cm³ glacial acetic (ethanoic) acid; 25 cm³ ethanol.

(ii) **Acetic orcein.** 2 g orcein; 100 cm³ glacial acetic (ethanoic) acid. Dilute a small portion with an equal volume distilled water just before use.

(iii) **Molar hydrochloric acid** (i.e. one mole of HCl per litre). Make up 87·3 cm³ of concentrated acid to 1 litre by adding distilled water.

(iv) **Glycerol albumen.** Shake 50 cm³ egg white with a few drops of dilute acetic acid; add 50 cm³ glycerol and 1 g sodium salicylate. Filter.

Setting up experiments. It is essential that the females used for the breeding experiments have not already mated with the males in the culture bottle. From a flourishing culture with many unhatched pupae, all the flies are shaken into a clean, dry bottle and the culture bottle re-plugged. The flies that emerge from the pupae in the culture bottle during the day will be virgins and can be easily recognized by the unexpanded wings. These flies are etherized, and males and females sorted into separate dry tubes where they can recover from the ether. About three virgin females of a given strain, e.g. wild type, are transferred to a fresh culture tube and six males of a different strain, e.g. vestigial wings, are introduced. The males need not be virgins and can be taken directly from stock cultures.

The flies will mate and lay eggs in the culture medium. Larvae hatch from the eggs, burrow through the food, grow, and in about 10–14 days from laying, pupate on the sides of the tube. The parent flies should be removed after one week. When the F₁ progeny have been emerging from the pupal cases for about 10 days they should be etherized, the different sexes and strains counted and recorded, and the flies killed in alcohol or retained for F₂ experiments. From the results, the ratios of the different strains can be calculated, and interpretations attempted on the lines of the principles of Mendelian inheritance.

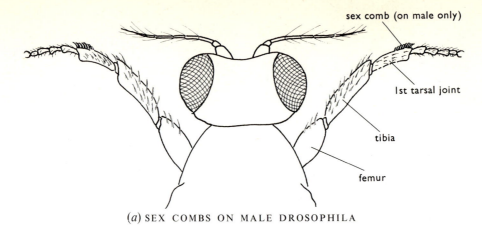

(a) SEX COMBS ON MALE DROSOPHILA

MALE more rounded, last 4 segments fully pigmented

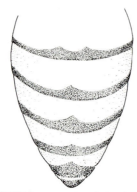

FEMALE more pointed, bands of pigment separated

(b) DROSOPHILA ABDOMEN: DORSAL ASPECT
These differences are not easy to see on newly hatched flies or ebony-body mutants.

Fig. 1.29   Distinguishing between male and female Drosophila

Sexing (see Fig. 1.29a and b). The presence or absence of sex combs on the fore legs is the surest guide. A hand lens or binocular microscope is essential.

Suitable crosses. Wild type × vestigial wing; wild type × ebony body; wild type × white eye: each cross should be made in two ways, reversing the sexes, e.g.

                vestigial winged male × wild type female
and   wild type male × vestigial winged female.

Etherizing (ether is very inflammable; no naked flame should be allowed while it is in use). The technique is depicted in Fig. 1.30a. The flies should not be exposed to ether for more than one minute. If they begin to recover while being counted, etc. they can be covered for a few seconds by a Petri dish lid carrying a small pad of ether-soaked cotton wool (Fig. 1.30b). If etherized flies are placed directly into a culture tube, the latter should have dry sides and be placed horizontally, otherwise the anaesthetized flies will stick to the food or the glass.

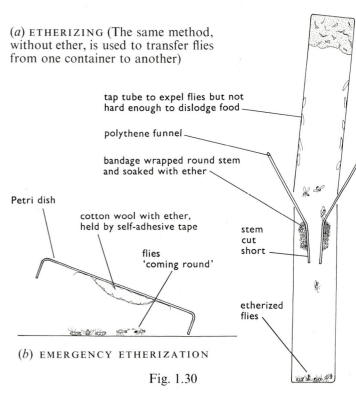

(a) ETHERIZING (The same method, without ether, is used to transfer flies from one container to another)

(b) EMERGENCY ETHERIZATION

Fig. 1.30

Further information

More detailed instructions can be obtained from the following books

Practical Heredity with Drosophila, by G. Haskell (Oliver & Boyd, 1961).

Biological Science: An Inquiry into Life (BSCS Yellow Version, published in Great Britain by Rupert Hart-Davis, 1963):
        Teachers' Manual for Students' Laboratory Guide, p. 193;
        Students' Laboratory Guide, p. 213.

Biology: An Environmental Approach, Man and his Environment (John Murray, 1972) p. 58.

Genetics for Schools (Modern Science Memoir No. 31), by Professor K. Mather (John Murray, 1953).

2 | Heredity and Genetics

FROM its parents an individual inherits the characteristics of the species, e.g. man inherits highly developed cerebral hemispheres, vocal cords and the nervous co-ordination necessary for speech, a characteristic arrangement of the teeth and the ability to stand upright with all its attendant skeletal features. In addition, he inherits certain characteristics peculiar to his parents and not common to the species as a whole, e.g. hair and eye colour, blood group and facial appearance. The study of the method of inheritance of these 'characters' is called genetics.

In sexual reproduction a new individual is derived only from the gametes of its parents. The hereditary material must therefore be contained in the gametes. For many reasons, this material is thought to be present in the nucleus of the gamete and located on the chromosomes (see pp. 6, 9 and 14).

Genes and inheritance

The term gene was originally applied to purely theoretical units or particles in the nucleus. These particles, in conjunction with the environment, were thought to determine the presence or absence of a particular characteristic. On p. 17 it was suggested that the genes may correspond to regions on the chromosomes and may consist of a large group of organic bases linked in a particular sequence in the chromosome.

In some cases, the presence of a single gene may determine the appearance of one characteristic, as in the eye colour of Drosophila (p. 8), but most human characteristics are controlled by more than one gene. This *multifactorial* inheritance and the impossibility with humans of breeding experiments, make it difficult to collect and present simple, clear-cut genetical information about man. In order to provide some clear ideas about heredity, simple cases amongst other animals will first be considered.

Single-factor inheritance. If a pure-breeding i.e. homozygous (*see* below) black mouse is mated with a pure-breeding brown mouse, the offspring will not be intermediate in colour i.e. dark brown or some combination of brown and black, but will all be black. The gene for black fur is said to be *dominant* to that for brown fur because, although each of the baby mice, being the product of fusion of sperm and egg, must carry genes for both blackness and brownness, only that for blackness is expressed in the visible characteristics of the animal. The gene for brown fur is said to be *recessive*. The black babies are called the first filial or F_1 *generation*. If, when they are mature, these F_1 black mice are mated amongst themselves, their offspring, the F_2 *generation* will include both black and brown mice and if the total number for all the F_2 families are added up, the ratio of black to brown babies will be approximately 3 to 1. It must not be assumed, however, that if two black F_1 mice have 4 babies, 3 will be black and one brown. In a mating which produced, say, 8 babies, it would not be at all unusual to find all black, or 5 black to 3 brown etc. The ratio 3:1 appears only when large numbers of individuals are considered.

The appearance of brown fur in the second generation is proof of the fact that the F_1 black mice carried the recessive gene for brown fur even though it did not find expression in their observable features.

In explanation, it will be assumed that a pure-breeding black mouse carries, on homologous chromosomes (*see* p. 6), a pair of genes controlling the production of black pigment. The genes are represented in subsequent diagrams (Fig. 2.1a and b) by the letters BB, the capital letters signifying dominance.

In the same position on the corresponding chromosomes in brown mice are carried the genes bb for brownness. The genes B and b are called *allelomorphic genes* or *alleles*. During the formation of gametes the process of meiosis (p. 11) will separate the homologous chromosomes, so that the gametes will contain only one gene from each pair. All the sperms from the pure-breeding black parent will carry the factor B and all the eggs from the brown parent will carry the factor b. When the gametes fuse, the zygotes will contain both factors B and b but since B is dominant to b, only the former gene is expressed, i.e. the offspring will all be black.

When, later on, these black F_1 mice produce gametes, the process of meiosis will separate the chromosomes carrying the B and b factors (*see* Fig. 2.1b) so that half the sperms of the male parent will carry B and half will carry b. Similarly, half the ova from the female will contain B and half b. At fertilization there are equal chances that a B-carrying sperm will fuse with either an egg carrying the B gene or an egg with the b gene so producing either a BB or a Bb zygote. Similarly there are equal chances of a b-carrying sperm fusing with either a B- or a b-carrying ovum to give bB or bb zygotes.

This results in the theoretical expectation of finding, in every four F_2 offspring, one pure-breeding black mouse BB, one pure-breeding brown mouse bb, and two "impure" black mice Bb.

The separation at meiosis of the alleles B and b into different gametes is called *segregation*. The pure-breeding black (BB) and brown (bb) mice are called *homozygous* for coat colour and the "impure" black mice (Bb) are called *heterozygous*. The heterozygous mice will not breed true, i.e. if mated with each other their litters are likely to include some brown mice. The homozygous BB mice mated together can produce only black offspring and the bb homozygotes only brown offspring.

Genotype and phenotype. The BB mice and Bb mice will be indistinguishable in their appearance, i.e. they will both have black fur and they are thus said to be the same *phenotypes*, in other words they are identical in appearance for a particular characteristic, in this case blackness. Their genetic constitutions, or *genotypes*, however, are different, namely BB and Bb. In short, the black phenotypes have different genotypes.

To distinguish between the black phenotypes, the usual practice is to do a further breeding experiment called a *back-cross*.

The back-cross. To discover their genotypes, the black F_2 phenotypes are each mated with mice of the same genotype as their brown grandparent i.e. the homozygous recessive, bb. Half the gametes from the heterozygous black mice Bb will

SINGLE-FACTOR INHERITANCE

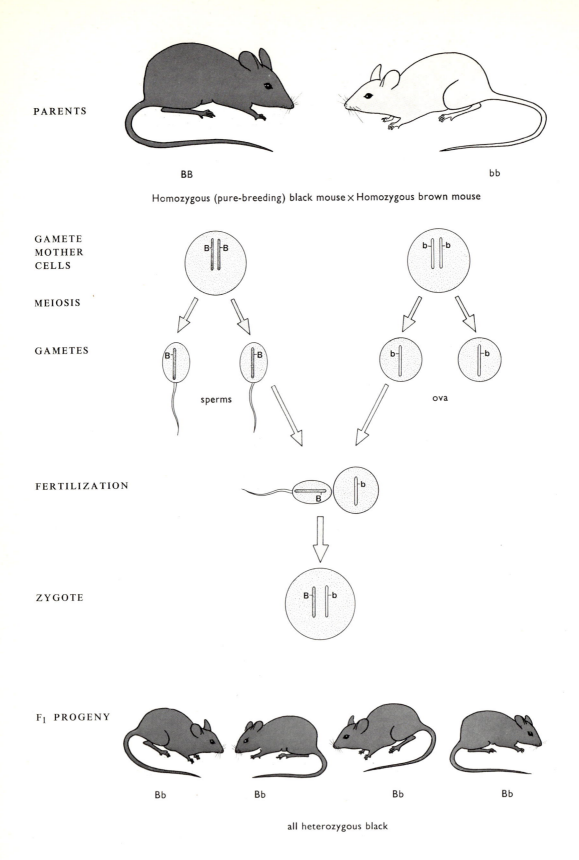

Fig. 2.1(*a*)   Inheritance of a single factor for coat colour in mice

SINGLE-FACTOR INHERITANCE

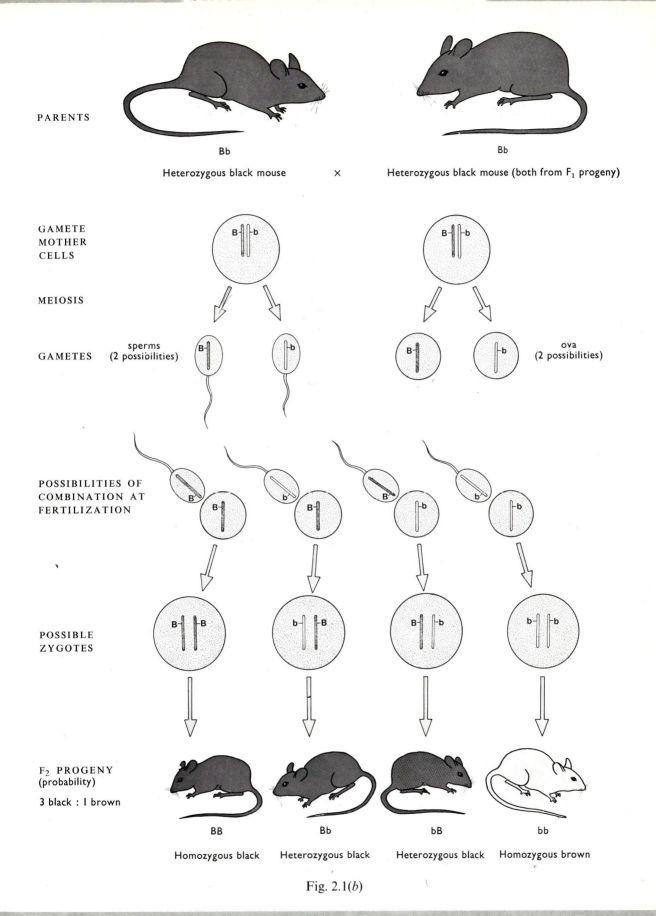

PARENTS

Bb
Heterozygous black mouse   ×   Bb
Heterozygous black mouse (both from F₁ progeny)

GAMETE
MOTHER
CELLS

MEIOSIS

GAMETES   sperms
                 (2 possibilities)   ova
                                                       (2 possibilities)

POSSIBILITIES OF
COMBINATION AT
FERTILIZATION

POSSIBLE
ZYGOTES

F₂ PROGENY
(probability)

3 black : I brown

BB
Homozygous black   Bb
Heterozygous black   bB
Heterozygous black   bb
Homozygous brown

Fig. 2.1(b)

carry the B gene and half will carry the b gene. The gametes from the homozygous black mouse, BB, will all carry the B gene. Similarly, the gametes of the homozygous recessive brown mouse will all carry the b gene. Thus, when the black parent is heterozygous, one would expect the back-cross to yield approximately equal numbers of black and brown babies in the litters but if the black parent was homozygous, the babies must all be black since they receive the dominant gene for blackness, B, from this parent (Fig. 2.2).

Incomplete dominance (*codominance*). If red Shorthorn cows are mated with white Shorthorn bulls the coats of the calves carry both red and white hairs, giving a *red roan*. Neither the red nor the white factor is dominant.

Black phenotype BB   ×   Brown "grandparent" bb

gametes   All B   All b

Offspring will all be Bb (black phenotypes)

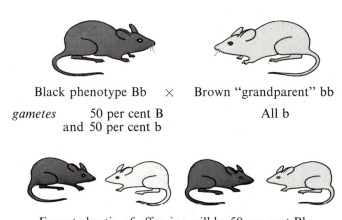

Black phenotype Bb   ×   Brown "grandparent" bb

gametes   50 per cent B   All b
and 50 per cent b

Expected ratio of offspring will be 50 per cent Bb
(black) and 50 per cent bb (brown)

Fig. 2.2   The back-cross

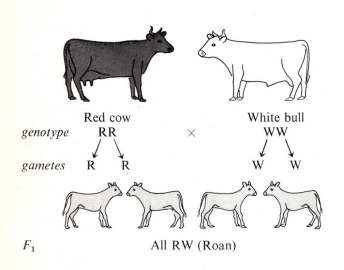

	Red cow		White bull
genotype	RR	×	WW
gametes	R R		W W

F_1 All RW (Roan)

Red cows and bulls when mated together will breed true, i.e. all their offspring will have red coats. White cows and bulls, similarly, are homozygous and will breed true. The F_1 roan cattle, however, are heterozygous and will not breed true; their progeny will include red calves and white calves as well as roans.

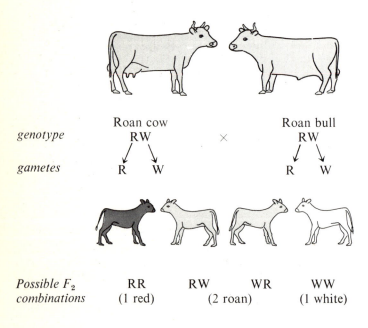

	Roan cow		Roan bull
genotype	RW	×	RW
gametes	R W		R W

Possible F_2	RR	RW	WR	WW
combinations	(1 red)	(2 roan)		(1 white)

The inheritance of the ABO blood groups in man is an instance of incomplete dominance. According to whether their blood will mix without clotting during a transfusion, people are classified into four major blood groups, A, B, AB and O. The blood group is controlled by three genes, A, B and O acting at the corresponding site on homologous chromosomes. A person will inherit two of these genes, one from each parent. Gene O is recessive to both A and B, but A and B are co-dominant, i.e. if a person inherits gene A from one parent and gene B from the other, he will be group AB, neither gene being dominant to the other. It follows that group O people must have the genotype OO while group A persons could be AA or AO and group B individuals BB or BO. The following example shows the possible blood groups of children born to a group A man and a group B woman both of whom are heterozygous for these genes.

Phenotype	group A		group B	
Genotype	AO		BO	
Gametes	A and O		B and O	
F_1 *Genotype*	AB	AO	OB	OO
Phenotype	group A B	group A	group B	group O

Two-factor inheritance. (Fig. 2.3*a* and *b*.) In cattle, uniform colouring (U) is dominant to spotted (u). If pure-breeding (homozygous) black bulls (BBUU) are mated with red spotted cows (bbuu), all the calves of the F_1 generation will be black and uniformly coloured. When two of the F_1 generation are mated, 4 types of offspring are possible as shown in Fig. 2.3*a* and *b*. With 4 possible types of gamete, there are 4^2 possible combinations in the zygote.

Although the calves carrying genes Bb Uu, BB Uu or Bb UU are heterozygous for one or both characters, they will be indistinguishable from BB UU calves as a result of the dominance of black and uniform genes. These calves have the same *phenotype*, i.e. the visible coat characteristics are the same, black and uniform, but the animals have a different genetic constitution or *genotype*.

Two new combinations have arisen, black and spotted Bb uu or BB uu, and red and uniform bb Uu or bb UU. These are called *recombinations*.

The expected ratio of combinations from a large enough number of matings would therefore be 9 black uniform, 3 black spotted, 3 red uniform and 1 red spotted. These results indicate that there is no linkage between the genes for colour and distribution of colour, i.e. they are presumed to be on different chromosomes.

Human genetics

The 'one gene–one character' effects described above illustrate very clearly the Mendelian* principles of inheritance, but they are the exceptions rather than the rule. Rarely do single genes control one trait. Colour in sweet peas, for example, is controlled by two pairs of genes, CC and RR. Gene C controls the production of the colour base and gene R the enzyme which acts on it to make a colour. The recessive cc will produce no colour base and rr will have no enzyme. CCrr and ccRR combinations will thus be unable to produce coloured flowers. At least six factors operate to produce coat colour in mice. In man, eight of the chemical changes involved in blood clotting are known to be under genetic control so that several genes are responsible for coagulation, and absence of any one of them may lead to a blood-clotting disease such as haemophilia (p. 16).

When one gene only is responsible for an important physiological change, its absence or modification will have serious consequences. Therefore most known instances of single-factor inheritance in man are associated with rather freakish abnormalities. These are usually rare conditions, e.g. occurring once in 10,000 to 100,000 individuals, but there are a great number of different kinds of genetic abnormality.

Examples of known single-factor inheritance involving a dominant gene in man are white forelock, woolly hair, one type of dwarfism, one form of night-blindness, and one form of *brachydactyly*, in which the fingers are abnormally short owing to the fusion of two phalanges. Recessive single factors are known to control, for example, red hair, inability to taste phenylthiourea, red-green colour vision, and one form of *albinism* which is the absence of pigment from the eyes, hair and skin.

In experimental animals or plants, the type of inheritance and the genetic constitution can often by established by breeding together the progeny of the F_1 generation, or by back-crossing one of the F_1 individuals with the mother or father and producing numbers of offspring large enough to give results that have statistical significance. These methods are obviously not applicable to man and our knowledge of human genetics comes mainly from detailed analyses of the pedigrees of families, particularly those showing abnormal traits such as albinism, from statistical analysis of large numbers of individuals from different families for characteristics such as sex ratio, intelligence, susceptibility to disease, etc., and from individual studies of identical twins (*see* below).

Despite the scarcity of evidence for single-factor inheritance in humans, there is plenty of evidence to suggest genetic control of many physiological, physical and mental characteristics. Body height, eye colour, hair colour and texture, susceptibility to certain diseases, and facial characteristics are all genetically controlled but in a more complex way than that described for mice on p. 23.

Heredity or environment?

Characteristics such as body size, health and intelligence are influenced by the environment. Genes for large body size are not likely to find expression in an individual deprived of adequate food and vitamins during his period of growth; lack of education will not provide a suitable background for the development of intelligence or intellectual activity even if the genes for it are present.

The birth of a mentally defective child may be the result of genic action or be due to the mother's having caught german measles in the first two months of her pregnancy. Traits such as eye colour or finger prints, on the other hand, are genetically determined and cannot be altered by the environment. The extent to which human characteristics are determined by heredity or environment is a fascinating problem on which there is little direct quantitative evidence and much controversy.

Identical twins

Twins may be either identical or fraternal. If they are fraternal, they are the result of the simultaneous fertilization of two separate ova by two separate sperms. The resulting zygotes will thus contain sets of chromosomes very different from each other as explained on pp. 12–14, and the twins, although they develop simultaneously in the uterus, will not necessarily be any more alike than if they were brothers or sisters born at different times, e.g. they can differ in sex. Identical twins, on the other hand, result when a single fertilized ovum, usually after a period of cell division, separates into two distinct embryos. The two embryos will thus have in their cells identical sets of chromosomes, since they are derived by mitosis (pp. 1–5) from a single zygote. The twins often share a placenta, although they may be enclosed in separate amnions. Such twins, having the same genotypes, usually resemble each other very closely and are invariably of the same sex, though variations in their position and blood supply while in the uterus may produce differences at birth.

Since the one-egg twins carry identical sets of genetic 'instructions', it can be argued that any differences between them are due, not to their genes, but to the effects of their environment. Identical twins, therefore, are a valuable source of evidence for assessing the relative importance of heredity and environment. For example, the average difference in height

---

* The term is derived from an Austrian monk, Gregor Mendel, who in the 1850s first discovered the type of inheritance described here.

TWO-FACTOR INHERITANCE

Fig. 2.3(*a*)   Inheritance of two factors for coat colour in cattle

TWO-FACTOR INHERITANCE

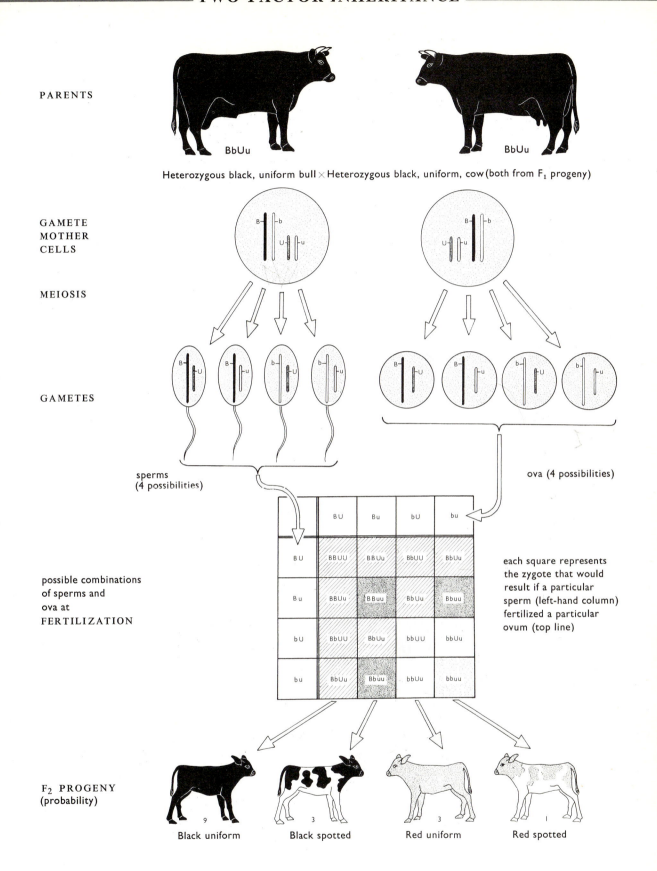

Fig. 2.3(b)

of 50 pairs of identical twins reared together was only 1·7 cm while the average difference for the same number of non-identical twins was 4·4 cm. These and other points of comparison are given in the table below:

TABLE 1

Average differences in selected physical characteristics between pairs of twins

Difference in:	50 pairs of identical twins reared together	50 pairs non-identical twins reared together	19 pairs of identical twins reared apart
Height (cm)	1·7	4·4	1·8
Weight (lb)	4·1	10·0	9·9
Head length (mm)	2·9	6·2	2·2
Head width (mm)	2·8	4·2	2·85

(From Freeman, Newman and Holzinger, *Twins: A study of heredity and environment*, Univ. Chicago Press, 1937)

The scores of identical twins in intelligence tests (Table 2) show a greater correlation than those of non-identical twins even when the former have been brought up in different environments, but the results show that educational background can make a considerable difference.

TABLE 2

Corrected average differences in I.Q. tests for 50 pairs of identical twins

Identical, reared together. 50 pairs	Non-identical. 52 pairs	Identical, reared apart. 19 pairs
3·1	8·5	6·0

(From Freeman, Newman and Holzinger, *Twins: A study of heredity and environment*, Univ. Chicago press, 1937)

Detailed histories of identical twins give even more impressive results. There is the case of girl twins, separated soon after birth; one was brought up on a farm and the other in a city, and both became affected with T.B. at the same age. Identical twin sisters were separated and adopted shortly after birth but both became schizophrenic within two months of each other in their 16th year. Such individual examples are of interest but cannot lead to any far-reaching conclusions about inheritance of human characteristics in general.

Applications of genetics to human problems

(a) **Eugenics** is the study of human genetics from the point of view (i) of encouraging breeding for good characteristics to improve human stock and (ii) of eliminating or reducing harmful characteristics. The simplest notions of eugenics as appreciated by the layman have tended to fall into some disrepute since they are based on naïve assumptions and insufficient evidence. It would seem obvious at first sight that if all the congenitally* insane were sterilized or prevented from breeding in some way, the numbers of congenital idiots in the world would be greatly reduced or the condition completely eliminated. That this is not the case is shown by considering the condition of recessive albinism. Only people with the two

recessive genes, aa, will show the characteristics. These number about one in 20,000 of the population. The marriage of two albinos would give rise to all albino children but such a marriage is unlikely. Calculations show that 1 in 70 of the general population must be a carrier for albinism, i.e. have the genetic constitution Aa. Such persons are indistinguishable, at the moment, from normal AA individuals but a marriage between two such people could produce some albino children aa. Of all albinos, 99 per cent arise from such marriages between unwitting carriers and clearly there could be no question of sterilization or voluntary birth control at least until after the first albino child was born and the genetic constitution of the parents so revealed. If no homozygous recessive persons, aa, were to breed, it would take 22 generations to reduce the frequency of a harmful gene from 1 per cent to 0·1 per cent.

Similarly, if every congenitally mental defective were sterilized, the incidence of the disease would fall by only 8 per cent. This fall, however, represents the elimination of suffering for a large number of people; mentally defective people can hardly provide a suitable home even if they have normal children.

The elimination of a harmful dominant trait, Dd or DD, can theoretically be accomplished in one generation provided that the condition appears before the reproductive age.

(b) **Genetic advice.** Even if the more drastic measures of sterilization, etc. are not adopted, advice based on sound knowledge of genetics can sometimes avoid unsatisfactory breeding and several countries have Genetic Counselling Services to give information about inborn defects.

For example, a person suffering from congenital juvenile cataract (a defect of the eye), caused by a dominant gene would be told that, if he married a normal person, half their children might inherit the disease; this can be seen by considering the possible F_1 offspring from the mating of Dd (affected) and dd (normal). An albino (aa) marrying a normal person who might be either AA or Aa would find that the chance of any one of his children being an albino is 1 in 140, while the normal brother of an albino, who might be a carrier, Aa, has a chance of 1 in 420 of begetting an albino child if he marries a normal person. The chances for other relations, such as aunts and uncles, can also be calculated.

Genetic advice could be far more accurate if heterozygous affected people could be distinguished from homozygous normal. In some hereditary diseases such a distinction is becoming possible. Heterozygotes for a disease called *phenylketonuria* can be detected with reasonable certainty by a test for the level of *phenylalanine* in their blood.

(c) **Consanguinity.** The study of human genetics enables predictions to be made on the chances of the recombination of two harmful, recessive genes in the children of marriages between first cousins.

Fig. 2.4 shows diagrammatically the theoretical pedigree of two cousins, Bill and Jane. Cousin Bill is assumed to be heterozygous for a recessive gene, Nn. (He would be known to be Nn for certain, only if one of his parents was nn.) Bill could have inherited this gene from his grandparents, A or B. If the gene came from grandparents B, there is a chance that Jane has inherited the gene from the common grandparent. There is thus a chance of 1 in 8 that cousin Jane is also Nn, in which

---

* In this context the term *congenital* is taken to mean a hereditary condition rather than an effect resulting from environmental causes during gestation or birth.

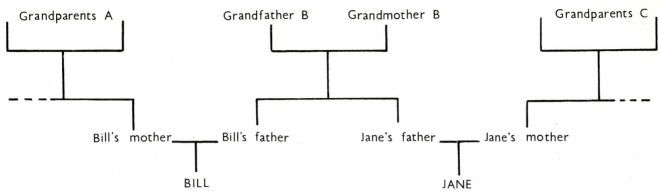

Fig. 2.4   Lineage of first cousins

case the chance of an affected child from their marriage is 1 in 4. The overall chances of an affected child are thus $1/8 \times 1/4 = 1/32$, i.e. 1 in 32.

If the gene is fairly rare in the general population, e.g. occurs once in 100 individuals, the chances of Bill marrying an Nn person from the general population are 1 in 100. The overall chances of an affected child in this case are $1/100 \times 1/4$, i.e. 1 in 400.

Although these considerations would apply equally well to beneficial genes, cousin marriages are not usually encouraged and brother-sister marriages forbidden by law. This does not reduce the total number of homozygous recessives which occur in a population but does reduce the chance of their occurring in a particular family. Consanguinity is bound to occur sooner or later, otherwise we should need to have had an impossibly large number of ancestors.

parents the mental equipment for intelligent thought, but this will not produce his full potential of intelligent behaviour unless he is educated.

That part of intelligence which is genetically controlled is almost certainly influenced by a large number of genes and is not susceptible to simple analysis. When a graph or histogram is plotted to show the different numbers of individuals possessing a particular I.Q. value, the type of picture obtained is that shown in Fig. 2.5, often called a 'normality' curve or *curve of normal distribution*. Similar curves are also obtained for factors such as height and skin colour and can be explained on the basis of several genes influencing the characteristic. Instead of the straightforward presence or absence of a condition, such as albinism, there is a continuous variation with every grade of intermediate.

Intelligence

Intelligence is a product of a person's genetic constitution and the effect of his environment; i.e. he may inherit from his

Discontinuous and continuous variation

The individuals within a species of plants or animals are alike in all major respects; indeed, it is these likenesses which determine that they belong to the same species. Nevertheless, even though an organism recognizably belongs to a particular species, it may differ in many minor respects from another individual of the same species. A mouse may be black, brown, white or other colours, the size of its ears and tail may vary but despite these variations, it is still recognizably a mouse.

When the presence or absence of a single gene produces clearly recognizable variations that cannot be affected by the environment, the variations are said to be *discontinuous*. The variations of coat colour for mice described on p. 23 are examples of discontinuous variation. In this case the mice are either black or brown, there are no intermediates. Discontinuous variations such as these cannot be altered by changing the diet or the temperature or other environmental factors. An achondroplastic dwarf cannot grow tall by eating more food; an albino cannot acquire a sun-tan because he is unable to produce melanin in his skin.

Examples of continuous variation are height, weight and intelligence. There are not merely two classes of people, tall and short, but a whole range of intermediate sizes differing by barely perceptible distances. The 'jumps' in the I.Q. levels of Fig. 2.5 result from the arbitrary choice of 9 units as the categories for comparison. Continuous variations, though genetically controlled, can be affected by the environment as explained on p. 27.

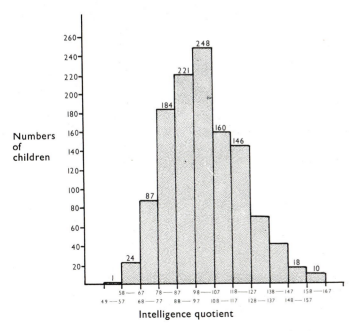

Fig. 2.5   Distribution of I.Q. rating in a random sample of 1,207 Scottish eleven-year-old children

(From C. O. Carter, *Human Heredity*, Penguin Books, 1962)

3 | Evolution and the Theory of Natural Selection

THE theory of evolution offers an explanation of how the great variety of present-day animals and plants came into existence. It supposes that life on Earth began in relatively simple forms which over hundreds of millions of years gave rise, by a series of small changes, to a succession of living organisms which became more varied and more complex. Taken to its logical conclusion, the evolutionary theory must suppose that life itself evolved from non-living matter. It must be emphasized that evolution is a theory and not an established fact. In general terms, it is an acceptable hypothesis to account for the existence of the living organisms which we know today.

The arguments for evolution are drawn from the kind of evidence which follows, though much of the evidence is either very incomplete or circumstantial.

1. Reproduction and spontaneous generation

As far as we know, all living organisms are derived by reproduction from pre-existing organisms and do not arise spontaneously from non-living matter. This knowledge weakens any alternative theory to evolution, if it claims that each kind of organism known today arose spontaneously or was created suddenly at different points in time, e.g. that horses were created suddenly one million years ago and have remained the same ever since, reproducing their kind exactly over thousands of generations.

The evolutionary theory would thus assume that when new forms of life appear on the Earth, they have been derived by reproduction from organisms which already exist, e.g. that mammals were derived from reptiles, reptiles from amphibia and amphibia from fish. In general, for vertebrates at least, the fossil record supports this contention. The question which must arise, however, is 'Where did the *first* living creatures come from?' To this the biologist is obliged to say that, although spontaneous generation of life from non-living substances is not known to occur today and is thought to be very improbable during the geological period of which we have some knowledge, there was a time, some 500 million years ago or more, when conditions were favourable for such an event or series of events. For example, the atmosphere might have been devoid of oxygen but rich in methane and ammonia which makes feasible the production of amino acids and proteins.

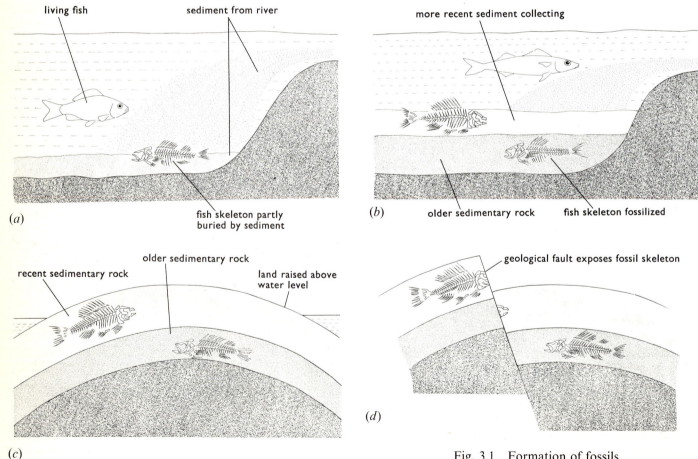

Fig. 3.1   Formation of fossils

32

2. The fossil record

Sedimentary rocks were formed by the settling down of mineral particles in lakes and oceans. These particles often became cemented together and the layers of sediment were compressed over millions of years to form rock. The dead remains of animals and plants, falling to the bottom of the lakes or seas became incorporated in the sediment and so preserved in a variety of ways as *fossils* (Fig. 3.1). As a result of slow earth movements, many sedimentary rocks were raised above the water and the fossils they contained became accessible for study.

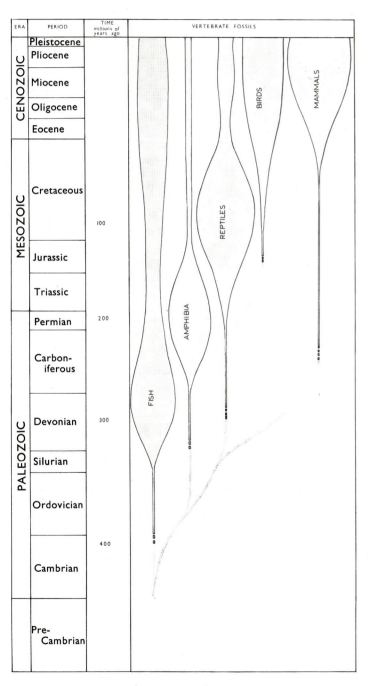

Fig. 3.2 Chart to show the earliest occurrence and relative abundance of fossil vertebrates (possible evolutionary relationships are shown by the faint lines)

(After Grove and Newell, *Animal Biology*, University Tutorial Press)

If not too contorted by the earth movements, in an exposed series the lowest sedimentary rocks will be the oldest, and from the fossils in successive layers, the scientist can form some idea of the animals and plants present millions of years ago. When the fossils in various layers are studied it appears (*a*) that many present-day animals and plants are not represented (Fig. 3.2), and (*b*) that a vast number of organisms represented by skeletal remains in the rocks no longer exist today. For example, 300 million years ago there were apparently no mammals, that is, no fossil remains have yet been found; but there were at that time some 'armour-plated' fish (Fig. 3.3) which no longer seem to exist.

Such evidence seems to detract from any idea that all the organisms existing today have been reproduced exactly since life began. Even if spontaneous generation or *biogenesis* could have taken place 300 million years ago, it seems unlikely that it would 'generate' anything so complex as a mammal in a single operation or even a number of operations in a very short time.

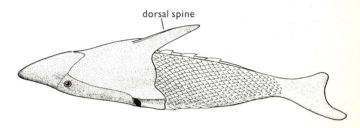

dorsal spine

Fig. 3.3 Pteraspis—one of the extinct, 'armour-plated' fish of the Silurian period (a reconstruction from fossil remains)

Alternatively it could be argued that mammals did exist 300 million years ago, but were so sparsely or locally distributed that we have not found any fossil remains. This does not help us to answer the question of how mammals arose, but simply pushes their hypothetical origin back to an earlier date.

More acceptable is the idea that mammals were derived from reptile-like ancestors by a long series of small changes, and the fact that scientists have found fossil remains of animals intermediate in many respects between reptiles and mammals lends support to this idea.

It must not be supposed, however, that mammals evolved from reptiles in the sense that present-day reptiles could produce a mammal, even over a very long period of time; or even that existing reptiles represent the form of the ancestral animals that gave rise to mammals. The evolutionary theory postulates that mammals and reptiles share one or a small number of common ancestors that were neither wholly reptilian nor wholly mammalian and which became extinct in due course. Similarly it supposes that reptiles had fish-like ancestors not represented amongst present-day fish and that each group has continued to evolve, but in different ways, since the first divergence.

There is little or no evidence in the fossil record to suggest that any of the large invertebrate groups of animals share a common ancestor. This may be because (*a*) our knowledge of the fossil record does not go back far enough, (*b*) the common ancestors, if they existed, have not been preserved, or (*c*) the events which produced living organisms from non-living matter occurred more than once, initiating a number of different, primitive forms of life that evolved into the invertebrates but share no common, living ancestor.

3. Circumstantial evidence

By studying the structure and distribution of modern animals it is possible to point out features which can be interpreted as supporting the theory of evolution. One such study is *comparative anatomy*, and perhaps the most familiar example is the skeleton of mammalian limbs (Fig. 3.4). The limbs of seals, moles, bats and antelope look very different from each other and are adapted to the functions of swimming, digging, flying and running. Despite this difference of appearance and function, they all have basically the same skeletal structure. If these mammals did not originate from a common ancestor but arose independently (and spontaneously?), there seems no convincing reason why the pattern of bones in limbs performing such different functions should be so similar. On the other hand, it does seem reasonable to visualize these limbs as modifications of the primitive, unspecialized limbs of a common ancestor, and that the differences came about during evolution as the limbs became more closely adapted to each special method of progression, while retaining the fundamental pattern of bones and joints.

Many other examples of comparative anatomy could be cited in the vertebrates, including fundamental similarities of the skeletal, circulatory and nervous systems. Also, other lines of circumstantial evidence, such as the similarity between the embryonic forms of different vertebrates, could be discussed. Being circumstantial evidence, i.e. attempting to fit known facts into a pattern of events in the past which is suspected but impossible to reproduce, the evidence is often subject to alternative interpretations and further discussion is suited to more advanced study than this book can offer.

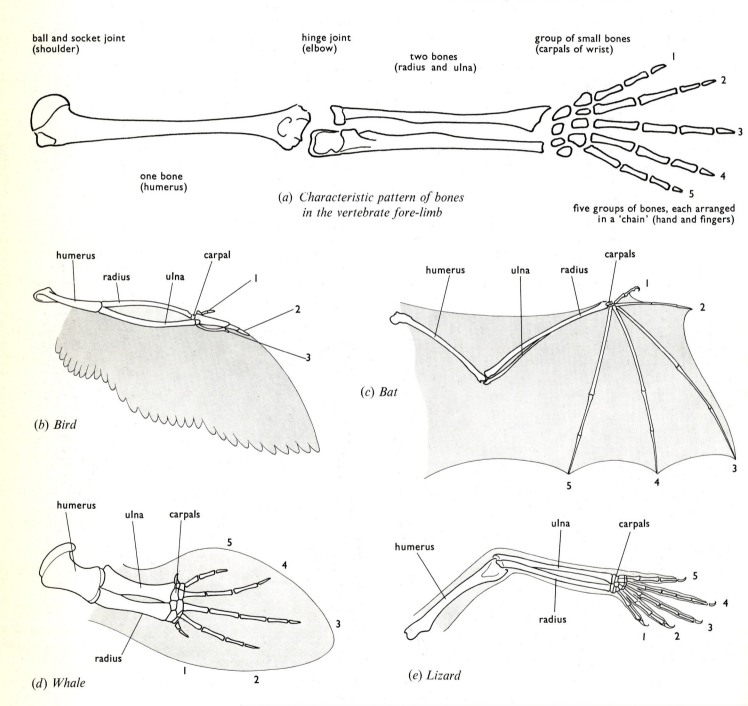

Fig. 3.4 Comparative anatomy of vertebrate limbs

NATURAL SELECTION

In 1858 Charles Darwin and Alfred Russel Wallace put forward a theoretical explanation of how evolution could have taken place and new species arisen. The theory of evolution by natural selection is the one which so far fits most, but not all, of the observed facts and has been strengthened by discoveries since 1858.

The arguments for the theory of natural selection can be summarized as follows:

(*a*) *Observation* 1. The offspring of animals and plants outnumber their parents.

(*b*) *Observation* 2. Despite this tendency to increase, the numbers of any particular species remain more or less constant.

(*c*) *Induction* 1. Since fewer organisms live to maturity than are produced, there must be a 'struggle' for survival.

(*d*) *Observation* 3. The individual members within any plant or animal species vary from each other by small differences; some of these differences can be inherited.

(*e*) *Induction* 2. (i) Some of these varieties are better adapted to the environment or mode of life of the organism and will tend to survive longer and leave behind more offspring. If the variations are harmful, the organism possessing them may die before reaching reproductive age and so the variation will not be passed on.

(ii) If an advantageous variation is inherited by an organism, it will also live longer and leave more offspring, some of which may also inherit the variation.

Small but favourable variations may thus accumulate in a population over hundreds of years until the organisms differ so much from their predecessors that they no longer interbreed with them. The 'variety' would now be called a new species. Certain points in the argument above need elaboration to make them clear.

(*a*) If a pair of rabbits had 8 offspring which grew up and formed 4 pairs, eventually having 8 offspring per pair, in four generations the number of rabbits stemming from the original pair would be 512, i.e. 2→8→32→128→512.

(*b*) An induction is an argument in which a generalization is made as a result of observations of actual events. The potential number for the fourth generation of rabbits is 512. If, however, the population of rabbits is to remain constant, 510 must have died or not been born, leaving only two of this vast potential family still alive.

The 'struggle' for survival, however, does not imply actual fighting; the participants may never meet but they could be in competition for food and shelter. Often the competition will not only be between adults, but between eggs or larvae or seeds, i.e. the most prolific stage of the life history of a species at which mortality rate is often high. The 'struggle' is often quite passive and may depend on the relative resistance of eggs to adverse conditions, or concealment patterns on the body leading to effective camouflage.

(*c*) Man is a species and the variations between members of this species are obvious at a glance. Variations in other organisms are less obvious but very evident to an experienced observer. If a variation is to play an effective part in evolution it must be heritable. A variation acquired in the lifetime of an organism, e.g. the well-developed muscles of an athlete, is not usually heritable.

(*d*) A species of a normally light-coloured moth, *Biston betularia*, the peppered moth, produces from time to time a black variety. This black variety was first recorded in 1848 in Manchester but by 1895 it had increased to 98 per cent of the population in this district. Observations showed that the light-coloured form was well camouflaged against the lichen-covered tree trunks where it normally rested. The atmospheric pollution of industrial areas, however, reduced or eliminated the lichens on the tree trunks and also darkened them with deposits of soot, so that the dark forms were better concealed (Plate 7) than

Plate 7. LIGHT AND DARK FORMS OF THE PEPPERED MOTH AT REST ON TREE TRUNKS
(From the experiments of Dr H. B. Kettlewell, University of Oxford)

(*a*) Soot-covered oak trunk near Birmingham   (*b*) Lichen-covered trunk in unpolluted countryside

the light forms. Better concealment led to fewer moths being eaten by birds, e.g. a Redstart in Birmingham was observed to eat 43 pale forms and only 15 dark forms from equal numbers resting on trees. The 'struggle' here is very indirect, being a 'struggle' for concealment, but the outcome is that more of the dark forms would survive and lay eggs. The dark colour in many cases is due to a single, dominant gene (p. 23) and so would be inherited by some of the offspring. The dark variety, however, is not yet a new species since it will interbreed with the light form, but this account illustrates how a new species could arise by natural selection.

HERITABLE VARIATION

The sources of heritable variation were not known to Darwin but have since been shown to arise principally in two ways: by mutation and by recombination.

Mutation and natural selection

On p. 19 it was explained that a mutation is a change which occurs spontaneously in a chromosome or gene.

Chromosome mutation may take the form of:

(*a*) *polyploidy*—the duplication of the entire set of chromosomes in the gametes or zygotes (Fig. 3.5).

(*b*) *translocation*—(i) a chromosome breaks and the fragment joins to a non-homologous chromosome. The total amount of genetic material is unaltered but completely new linkage groups are formed (Fig. 3.6);

(ii) centric fusion, in which two chromosomes with terminal centromeres join together to function as one chromosome.

(*c*) *aneuploidy*—one or more chromosomes are duplicated or missing. From an evolutionary point of view, polyploidy and centric fusion produce variations on which selection is most likely to act since aneuploidy, such as Down's Syndrome (p. 19), often results in failure to reproduce.

Centric fusion occurs naturally in a number of animal species, giving rise to variations which may have a selective advantage. A marine snail, similar to a dog whelk, which is common on the Pacific coast of North America has a haploid complement of 18 chromosomes but sometimes up to 5 of these are joined at the centromeres, so giving a chromosome count varying from 13 to 18. The 18-chromosome individuals flourish below the low-tide mark, the 13-chromosome forms

Cross-pollination between radish and cabbage plants produces F_1 hybrid plants which form only sterile seed pods because the two sets of 9 chromosomes cannot pair and separate properly at meiosis. If the chromosome set of the hybrids doubles, however, it produces a fertile tetraploid. This plant can set seed because the homologous chromosomes can now pair at meiosis and produce viable gametes.

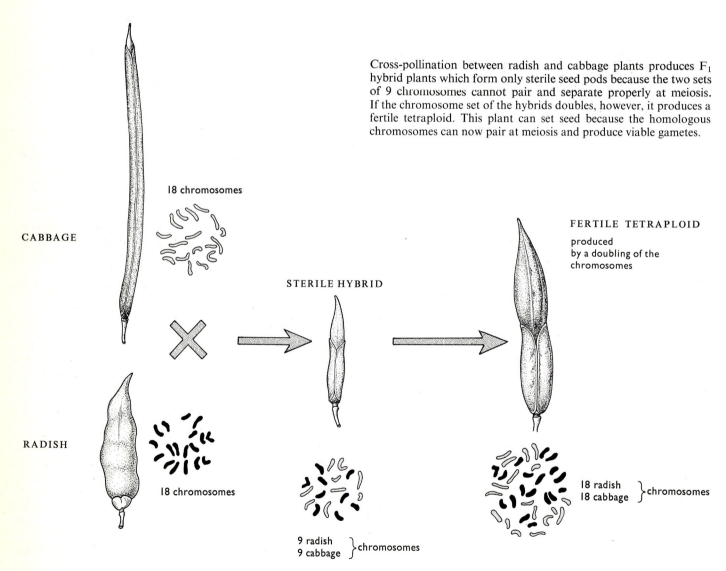

Fig. 3.5   Seed pods of cabbage and radish showing the effect of chromosome doubling

(After Karpechenko)

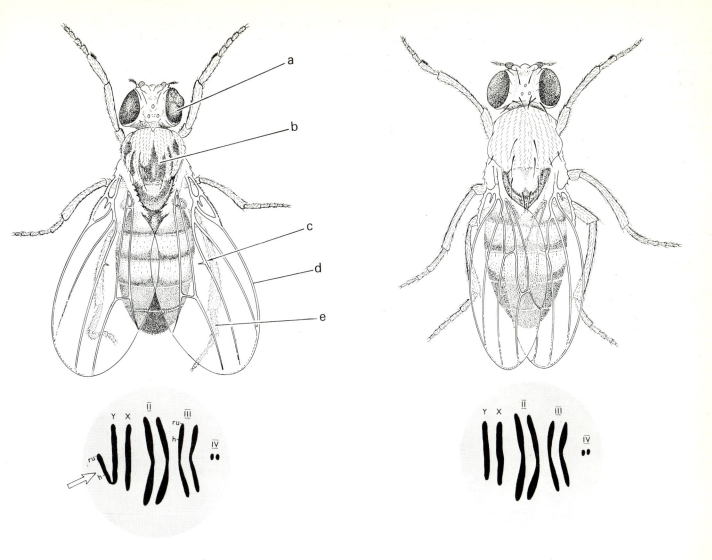

The cells of the fruit fly on the left have an extra segment of chromosome no. 3 which has become attached to the Y chromosome. As a result, this fly has (*a*) mis-shaped eyes, (*b*) dark patterned thorax, (*c*) imperfect cross veins, (*d*) broad wings, (*e*) incurved hind legs. The normal fly is shown on the right. (*See also* Fig. 1.18, p. 15.)

Fig. 3.6   Translocation   (After Muller, *Journal of Genetics*, 1930)

predominate near the high-water mark, and the others are found in between.

One of the evolutionary steps from ape-like ancestor to man might have involved centric fusion, since the gorilla, chimpanzee, and orang-utan have a diploid number of 48 chromosomes compared with man's 46.

Polyploidy, as illustrated in Fig. 3.5, can result in quite large evolutionary steps. A marine grass, *Spartina townsendii*, first recorded in 1870, was found to have a diploid chromosome number of 122. It had arisen probably from a cross between the British species *S. maritima* (60 chromosomes) and *S. alterniflora* (62 chromosomes) introduced from America. The doubling of the chromosome complement 2(30 + 31) gave a fertile hybrid which is more successful in colonizing tidal mud flats that is *S. maritima*. It seems likely that many present-day species of plants and animals have arisen by polyploidization from pre-existing forms. Cultivated plants in particular often have chromosome numbers which are multiples of those in their wild relatives.

Gene mutations occur when a section of the DNA in a chromosome is not copied exactly at cell division. For example, part of the haemoglobin molecule consists of a sequence of eight amino acids:

valine—histidine—leucine—threonine—proline
—*glutamic acid*—glutamic acid—lysine

This order will have been determined by the sequence of the four bases on part of a DNA strand (*see* p. 17). In persons suffering from a disease called sickle cell anaemia one of the DNA bases has not been correctly paired prior to the meiotic division which formed the gamete from which the individual developed. As a result, the sixth amino acid directed into the haemoglobin fragment depicted above is *valine* instead of glutamic acid. This small difference so alters the properties of the haemoglobin that in low concentrations of oxygen it becomes relatively insoluble and forms rod-like particles which distort and eventually destroy the red cells (Plate 8). The production of the 'faulty' DNA is a mutation and since this

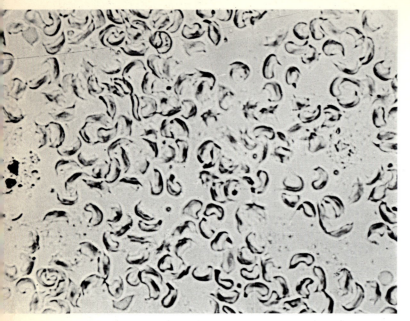

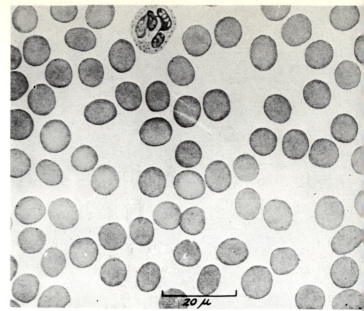

(Wellcome Museum of Medical Scien...)

Plate 8. SICKLE CELL ANAEMIA

(a) At low oxygen concentrations the cells become distorted   (b) Normal red cells for comparison

DNA will accurately reproduce itself in all the cells of the body, including the sex cells, the mutated gene will be passed on to the offspring. This is an example of a harmful mutation which is heritable. Natural selection operates on the phenotypes possessing the gene in the way described later, on p. 39.

Harmful mutations. Consider the following sequence of biochemical reactions which occur in cells of the liver:

in the urine which darkens on exposure to air, a symptom of *alcaptonuria*. Alcapton also collects in the cartilage of the joints causing a form of arthritis.

In the absence of enzyme '*c*', the production of melanin from 3,4-dihydroxy-phenylalanine is not possible, in which case the person will be an albino, lacking pigment in his skin, hair, and eyes.

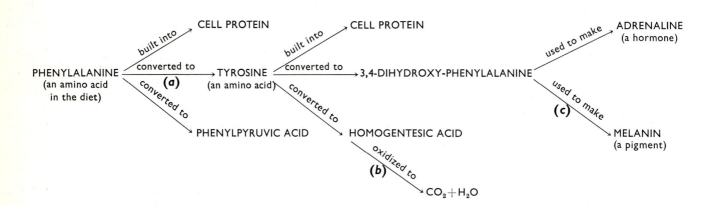

The letters '*a*', '*b*', and '*c*' represent enzymes which control the reactions indicated by the arrows. Each enzyme is a protein and is built up from amino acids according to the pattern determined by the base sequence of DNA in a gene. Mutations may occur in the genes which result in the failure to produce one or other of these enzymes.

If enzyme '*a*' is lacking, tyrosine cannot be made from phenylalanine. Since tyrosine is available from the normal diet this does not matter but blocking this pathway leads to excessive production of phenylpyruvic acid. This compound collects in the cerebrospinal fluid and is thought to cause the brain damage leading to mental retardation characteristic of the disease *phenylketonuria*.

If there is no enzyme '*b*', homogentesic acid (alcapton) is not oxidized to carbon dioxide and water but excreted unchanged

The three mutated genes causing these defects are recessive to the normal genes so that the clinical condition is expressed only in individuals homozygous for the gene.

When an organism is well adapted to a stable environment, it is not surprising that a mutation which disturbs its metabolism is harmful. When an environment is changing, however, a mutation which is harmful in one situation may be beneficial in another. For example, the mutation which produces a dark moth in a Dorset wood will be selectively harmful because the moth is more conspicuous to predators (Plate 7, p. 35). The same mutation in Birmingham, however, is selectively advantageous in a woodland environment devoid of lichen-covered tree trunks.

Mutation in bacteria. Mutations can occur whenever DNA is reproducing itself, an event which takes place prior to cell division. Some bacteria can divide every 25 minutes and it follows that there is a very high chance of mutations occurring per unit of time. One such mutation produces a resistance to specific drugs and occurs about once in every thousand million cell divisions. This seems very infrequent but a colony of ten bacteria could produce a population of that order in about nine hours, and it is likely that a thriving colony will contain several individuals resistant to, say, penicillin. In the 'normal' environment the mutation may have no advantage but in the presence of penicillin, all but the mutants will be destroyed. The progeny of the surviving mutants will inherit the resistance to penicillin and in this way a population of resistant bacteria can soon become established. For this reason, the widespread use of antibiotics is discouraged.

This example hardly seems like 'natural' selection because it implies that mutations conferring resistance to drugs occur even before the drug has been discovered. The argument which tries to explain this 'anticipatory' mutation points out that most antibiotics are derived from micro-organisms, such as fungi and bacteria, which inhabit the soil and probably compete with each other for the available food. It is possible that the production of antibiotic chemicals such as penicillin from the mould *Penicillium* plays a part in this competition, suppressing the activity of bacteria which could deprive the mould of its food. A mutation in the bacteria, making them resistant to penicillin, would have considerable selective advantage in this situation.

If the bacteria which are harmful to man have been derived by evolution from those in the soil, it seems plausible that they will have retained the genetic constitution which still, on rare occasions, mutates in such a way as to confer resistance to the antibiotics produced by their one-time competitors.

It must be admitted, however, that it is difficult to demonstrate that those organisms which produce antibiotics when grown in artificial media also do so in normal soil.

Significance of mutation. Since so many mutations are harmful to the individual bearing them and are eliminated by natural selection it might seem that a well-adapted organism with no mutations would be at an advantage. This may be true so long as the environment does not alter and the organism does not move to a different situation where it is subjected to new selection pressures for which it is poorly adapted. If a species of bacteria were incapable of mutation, exposure to streptomycin might eliminate the species. The fact that streptomycin-resistant mutants occur, allows the bacteria to survive such an adverse change in the environment.

The dark mutant of the peppered moth had little selective advantage prior to the Industrial Revolution but with the advent of pollution and the consequent darkening of tree trunks, the mutation enjoyed favourable selection pressure. The elimination of mutants by natural selection in one environment is the price which populations have to pay if they are to retain the ability to adapt to a changing environment. Mutations should not therefore be regarded as 'accidents' since they are events which occur normally in all populations and are essential for survival and evolution, though they may well be harmful to the many individuals in which they occur.

Mendelian recombination and natural selection

A mutant gene which is harmful in one genetic constitution (genotype) may be neutral or beneficial in another. Similarly, certain advantageous genes may exist in one population and different beneficial genes in another. If interbreeding takes place between the populations, individuals may arise in which the two beneficial genes recombine, so conferring a selective advantage on the new phenotype. For example, the genes in a wild grass which render it resistant to fungus disease have been combined by cross-breeding with the genes for large grain size in cultivated wheat thus producing a variety with a high yield and good resistance to disease. Although segregation (p. 23) tends to separate these beneficial genes their combined selective advantage could cause them to maintain a constant frequency in a population as happens with sickle cell trait.

Sickle cell anaemia occurs when two recessive mutant genes controlling haemoglobin production, combine in an individual. Such individuals have severe anaemia and generally die before reaching the reproductive age, i.e. natural selection removes the 'h' genes from the population. The recessive genes remain in the population, however, in the heterozygous individuals, e.g. HH=normal; hh=sickle cell anaemia; Hh=heterozygous for sickle cell anaemia. Although from one quarter to one half of the haemoglobin of the heterozygotes is affected, usually less than 1 per cent of their red cells show sickling in low oxygen concentrations. The heterozygotes are said to have the sickle cell *trait*. When two heterozygotes marry, one would expect on average a quarter of their children to have sickle cell anaemia

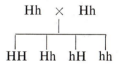

Hh × Hh

HH Hh hH hh

and would die before reaching reproductive age. In this way the h genes would be eliminated by selection from the population since for every four offspring from the HH genotypes there will be only three from the Hh genotypes. In certain African populations, however, as many as 34 per cent of the people carry the recessive gene and there is evidence to suggest that this is because the heterozygotes are more resistant to malaria, i.e. persons with the genotype Hh enjoy a selective advantage in malarious districts while the HH genotypes are reduced by the selective pressure of malaria. Investigations of Negro populations in America show an incidence of only 4 to 5 per cent of the trait compared with the 15 to 20 per cent characteristic of the African population from which the migrants were originally derived. In some non-malarial regions therefore, it looks as if the heterozygotes lose their selective advantage.

Balanced polymorphism

The existence of genetically determined varieties within a population provides the raw material on which natural selection acts, with one or other variety being favoured or reduced by the selective process. However, in most populations over a short term, the different varieties persist in about the same numbers; for example, although in certain environments the light forms of some moths have a selective advantage, dark mutants still occur with a low but consistent frequency partly as a result of new mutations appearing in the population and partly because the dark forms enjoy some positive selective advantage. It seems that although the dark forms are more easily seen when resting on trees, they are better concealed from predators while in flight, thus selection operates in their favour on certain occasions.

The term 'balanced polymorphism' refers to the persistence of such varieties in a population. The marine snails mentioned on p. 36 show just such a balanced polymorphism as a result of the centric fusion of some of their chromosomes.

Variations in eye and hair colour and the existence of different blood groups are examples of genetically controlled variations in human populations. It is not always possible to discern the selective advantage of one or other variant but this does not mean that the variant is selectively neutral. Human blood group A would appear to have no selective advantage until statistical analysis reveals that such individuals are less affected by duodenal ulcers, a factor which may enhance their reproductive capacity. This is also an illustration of the fact that the characteristic which we recognize as being controlled by a gene or group of genes is not necessarily the one on which selection is acting.

Isolation and the formation of new species

A mutation or recombination may give rise to a variety with characteristics which enable it to colonize new areas not accessible to the parent stock. In this way the variety may form a breeding population which is isolated from the original population. Further mutations may occur in this isolated population and accumulate until individuals are incapable of interbreeding with the original stock. In this way a variety will have given rise to a new species.

There are many ways in which populations can become isolated. Geographical isolation is an obvious case where rivers, oceans or mountains separate populations. Another cause could be a difference in breeding season or incompatibility of mating behaviour. If variety *A* breeds only in April while variety *B* breeds in July, the breeding populations are effectively isolated even if they occupy the same area.

Evolution by natural selection can be visualized as proceeding as follows: mutations and genetic recombinations arise in a population; those which are not so harmful as to be lethal give rise to a balanced polymorphism; environmental change or migration of the population favours certain varieties which leave more offspring which inherit these same variations; isolation allows other favourable genes to accumulate in the population until it differs so much from the parental stock that interbreeding is impossible.

'Survival of the fittest'

This is an expression often used to summarize the theory of natural selection. In this context it must be emphasized that 'fit' refers not to health but to the production of a large number of offspring which survive to reproductive age. A person may be physiologically 'fit', i.e. healthy and strong, but if he fails to have any children he is less fit in the evolutionary sense than a physically weak and unhealthy individual who nevertheless has a large family. Generally speaking, an organism which is healthy and well adapted to its surroundings will survive for a long time and, reproducing each year of its life, will leave behind many offspring which inherit its advantageous characters. In human populations, natural selection can be said to operate on a characteristic only if it affects reproductive capacity. For example, achondroplastic dwarfs are less fit because they have, on average, only 0.25 child per parent compared with 1.27 children born to their normal brothers or

sisters. Fewer dwarfs marry, and childbirth is hazardous for the females. The gene for dwarfism being dominant would therefore tend to disappear from the population were it not maintained by mutation.

From an evolutionary point of view it is better to consider the fitness of a population of organisms than the fitness of an individual. Although, in the short term, fitness is dependent on the number of viable offspring, in the long term it depends also on the potential for change and adaptation in the population. There will be a great deal more variation for selection to act on within a population of organisms than there is in a single creature and its progeny.

Natural selection in man

There is no reason apparent at the moment why natural selection should not have been responsible for the evolution of man from an ape-like ancestor and there is no reason why selection should not still be acting on human populations. The incidence of the sickle cell trait is likely to be the result of selection pressure. Where there is heritable variation and environmental pressure, evolution by natural selection is possible.

However, with the growth and spread of modern medicine, public health measures, education, and other benefits of civilization, natural selection has ceased to play so immediate a role in human populations as it does in other organisms. Weakly babies who would die in 'natural' conditions can be saved, and the congenitally blind, deaf, and lame are enabled to live and reproduce. Some people see in this artificial preservation a tendency for mankind to evolve into a race of weaklings. This is to suppose, however, that one day we shall be without the benefits of modern medicine, good food, surgery, etc. It also wrongly supposes that every weakly child will grow into an enfeebled adult and have unhealthy children. Natural selection is a wasteful process and would have eliminated our diabetics, our short sighted, and our Rhesus babies, many of whom may have an important part to play in human affairs.

Man has devoted much thought and effort to minimizing the adverse effects of environmental change by providing himself with, for example, clothing, housing, central heating, air conditioning, an agricultural industry, and protection from predators and competitors. This does not mean that selective pressures have disappeared altogether, but they have changed. Resistance to diseases of modern society, indifference to crowding and noise, ability to withstand the pace of modern life or assimilate a modern diet without developing ulcers, may all have selective advantages if they are controlled to some extent by genes and are manifest before the end of normal reproductive age.

However, in most advanced countries selection will operate largely through parental decisions on how many children they will have, since it may be assumed today that nearly all the children will reach reproductive age. The 'fittest' will still be those with the largest families but this will depend on the conscious decision to procreate rather than on the number left after miscarriage, infant mortality, and disease have taken their toll.

Further Reading

Beryl Ashton, *Genes, Chromosomes and Evolution* (Longmans, 1967)

Isaac Asimov, *The Genetic Code* (John Murray, 1962)

S. A. Barnett (Ed.), *A Century of Darwin* (Heinemann/Mercury, 1962)

C. O. Carter, *Human Heredity* (Penguin, 1970)

C. D. Darlington, *Genetics and Man* (Penguin, 1966)

T. Dobzhansky, *Mankind Evolving* (Yale University Press, 1962)

W. H. Dowdeswell, *The Mechanism of Evolution* (Heinemann, 1963)

L. C. Dunn, *Heredity and Evolution in Human Populations* (Oxford, 1959)

Wilma George, *Elementary Genetics* (Macmillan, 1964)

Peter Kelly, *Evolution and its Implications* (Mills & Boon, 1962)

G. A. Kerkut, *Implications of Evolution* (Pergamon, 1960)

Lewis and John, *The Matter of Mendelian Heredity* (Churchill, 1964)

K. Mather, *Genetics for Schools* (John Murray, 1953)

P. B. Medawar, *The Future of Man* (Methuen, 1959)

John A. Moore, *Heredity and Development* (Oxford, 1963)

David Paterson, *Applied Genetics* (Aldus, 1969)

L. S. Penrose, *Outline of Human Genetics* (Heinemann, 1959)

Scientific American, 'Facets of Genetics' (W. H. Freeman, 1969)

Sinnott, Dunn and Dobzhansky, *Principles of Genetics* (McGraw Hill, 1958)

Curt Stern, *Principles of Human Genetics* (W. H. Freeman, 1960)

Swanson, Mertz and Young, *Cytogenetics* (Prentice Hall, 1967)

Index

Bold figures indicate where subject is defined or given main treatment